ADVANCES IN SPACE RESEARCH

The Official Journal of the Committee on Space Research (COSPAR)
A Scientific Committee of the International Council of Scientific Unions (ICSU)

VOLUME 22, NUMBER 12

LIFE SCIENCES: BIODOSIMETRY, CHROMOSOME DAMAGE AND CARCINOGENESIS

Symposium F2.4
Sponsors

NATIONAL AERONAUTICAL AND SPACE ADMINISTRATION (NASA)
EUROPEAN SPACE AGENCY (ESA)
NATIONAL SPACE DEVELOPMENT AGENCY OF JAPAN (NASDA)
RUSSIAN SPACE AGENCY (RSA)
COMMITTEE ON SPACE RESEARCH (COSPAR)

Program Committee

T. C. Yang, U.S.A.
G. Horneck, Germany
J. Lucas, U.S.A.
P. Rettberg, Germany
M. Durante, Italy
S. B. Curtis, U.S.A.
M. Mei, China

LIFE SCIENCES: BIODOSIMETRY, CHROMOSOME DAMAGE AND CARCINOGENESIS

Proceedings of the F2.4 Symposium of COSPAR Scientific Commission F which was held during the Thirty-second COSPAR Scientific Assembly, Nagoya, Japan, 12–19 July, 1998

Edited by

T. C. YANG

NASA Johnson Space Center, Mail Code SD23, Houston, TX 77058, U.S.A.

Published for

THE COMMITTEE ON SPACE RESEARCH

PERGAMON

U.K.	Elsevier Science Ltd, The Boulevard, Langford Lane, Kidlington, Oxford OX5 1GB, U.K.
U.S.A.	Elsevier Science Inc., 655 Avenue of the Americas, New York, NY 10010-5107, U.S.A.
JAPAN	Elsevier Science K.K., 9-15 Higashi-Azabu 1-chome, Minato-ku, Tokyo 106, Japan

First edition 1998

ISBN: 9780080436401

In order to make this volume available as economically and as rapidly as possible the author's typescript has been reproduced in its original form. This method unfortunately has its typographical limitations but it is hoped that they in no way distract the reader.

Printed and bound in Great Britain by CPI Antony Rowe, Chippenham and Eastbourne

CONTENTS

Pergamon

Adv. Space Res. Vol. 22, No. 12, p. 1627, 1998
Published by Elsevier Science Ltd on behalf of COSPAR.
Printed in Great Britain
0273-1177/98 $19.00 + 0.00

PII: S0273-1177(99)00134-9

EDITORIAL COMMENT

This is the first issue of *Advances in Space Research* devoted to the papers presented at the 32nd Scientific Assembly of COSPAR held in Nagoya, Japan, July 1998.

Raising the publication standards of *Advances in Space Research* has been a principal concern of the COSPAR Bureau and Publications Committee. With the 31st Scientific Assembly of COSPAR in Birmingham, UK in 1996, new procedures on refereeing and publishing were implemented. Editors were requested to have two referees for each manuscript. For the most part, the refereeing process was conducted after the assembly so that reviewers could devote the proper time and attention to each manuscript without additional obligations during the Assembly itself. New refereeing forms were introduced in keeping with other international scientific journals.

These efforts were intended to make the papers more worthwhile and result in scientifically superior volumes. While this new refereeing process delayed the publication of some of the Birmingham Assembly papers, the COSPAR Publications Committee believes that the quality of *Advances in Space Research* has significantly improved.

The next concern of the COSPAR Publications Committee is the timeliness of the volumes. In a letter to editors for the Nagoya Assembly, the COSPAR president requested that the editorial work be completed in a timely manner and all editors submit their papers to the publisher as quickly as possible. As before, all papers must be refereed by two reviewers. Papers received after each Editor's deadline will be published as Appendices to future volumes.

Dr Tracy Yang of the NASA Johnson Space Center volunteered to be editor of this COSPAR session, and initiated the editorial process shortly after the COSPAR Assembly. Unfortunately Dr Yang passed away in October 1998 with his editorial work partially completed. Shortly after Dr Yang's death, Dr T. K. Hei and Dr M. Durante assumed the responsibility of completing the editorial work; the complete package was submitted to the publisher in January, 1999.

Drs Hei and Durante are to be commended for completing this issue under extremely difficult circumstances. In a message to me, Dr. Hci stated that if one agrees to take up the editorship of a particular session, he or she should have some sense of responsibility of finishing the job in a timely manner. These two gentlemen certainly have lived up to these ideals. I thank both of them very sincerely for their efforts.

M. A. Shea
Editor-in-Chief

Pergamon

Adv. Space Res. Vol. 22, No. 12, p. 1629, 1998

Printed in Great Britain
0273-1177/98 $19.00 + 0.00

PII: S0273-1177(99)00027-7

Preface

This session of the 32nd Scientific Assembly of COSPAR focuses on biodosimetry and the genotoxic effects of heavy ion radiation in various species, including plants, rodent cells, and human cells.

Experimental results, based upon human serum vitamin D levels as well as chromosomal aberrations in circulating blood lymphocytes, are proposed biomarkers for both UV and heavy ion exposure. Molecular alterations in human epithelial cells transformed by heavy ions provide potential mechanisms for radiation induced carcinogenesis. Neoplastic transformations, based upon rodent cell models, provide consistent RBE-LET analysis.

The main scientific organizer is Dr. Tracy Yang, NASA Johnson Space Center, with Dr. Gerda Horneck, DLR German Aerospace Center, as deputy organizer. Doctor's Tom K. Hei of Columbia University and Marco Durante of NIRS at Chiba are the two reviewers for this scientific session. The organizers and the program committee would like to thank all those who contributed to make this symposium a success.

All contributing authors are deeply saddened by the untimely death of Dr. Tracy Yang, the editor and main scientific organizer of this session, three months after the meeting. The two reviewers subsequently completed the editorial task of this proceeding and dedicate this issue to the fond memory of Dr. Yang, a highly respected scientist and a wonderful human being.

Pergamon

Adv. Space Res. Vol. 22, No. 12, pp. 1631–1641, 1998

Printed in Great Britain
0273-1177/98 $19.00 + 0.00

PII: S0273-1177(99)00028-9

BIOLOGICAL MONITORING OF RADIATION EXPOSURE

G. Horneck

German Aerospace Center (DLR), Institute of Aerospace Medicine, Radiation Biology, D-51170 Köln, Germany

ABSTRACT

Complementary to physical dosimetry, biological dosimetry systems have been developed and applied which weight the different components of environmental radiation according to their biological efficacy. They generally give a record of the accumulated exposure of individuals with high sensitivity and specificity for the toxic agent under consideration. Basically three different types of biological detecting/ monitoring systems are available: (i) intrinsic biological dosimeters that record the individual radiation exposure (humans, plants, animals) in measurable units. For monitoring ionizing radiation exposure, in situ biomarkers for genetic (e.g. chromosomal aberrations in human lymphocytes, germ line minisatellite mutation rates) or metabolic changes in serum, plasma and blood (e.g. serum lipids, lipoproteins, lipid peroxides, melatonin, antibody titer) have been used. (ii) Extrinsic biological dosimeters/indicators that record the accumulated dose in biological model systems. Their application includes long-term monitoring of changes in environmental UV radiation and its biological implications as well as dosimetry of personal UV exposure. (iii) Biological detectors/biosensors for genotoxic substances and agents such as bacterial assays (e.g. Ames test, SOS-type test) that are highly sensitive to genotoxins with high specificity. They may be applicable for different aspects in environmental monitoring including the International Space Station.

INTRODUCTION

Radiation, both of ionizing and non-ionizing nature, is a natural element of the Earth's environment and has decisively shaped the evolution of life throughout its history. In response to environmental radiation encountered, life has developed a variety of defense mechanisms, such as increased production of stress proteins, activation of the immune defense system, and efficient repair systems for radiation-induced DNA injury. The concern about potential hazards from exposure to radiation results mainly from man-made radiation sources with long-lasting and sometimes unpredictable consequences to human health and ecosystem. Examples of such anthropogenic sources of ionizing radiation include medical applications and radiopharmacy, environmental radon exposure, nuclear weapons, and nuclear power technology. Likewise, the level of solar ultraviolet B (UV-B) radiation (280-315 nm) reaching the surface of the Earth increases as a consequence of the depletion of stratospheric ozone which is predominantly caused by man-made CFCs (chlorinated fluorocarbons) (Rowland, 1989). Furthermore, with the development of manned space flight, humans have started to leave the protective blanket of our atmosphere. In low Earth orbit (LEO), which is typical for most manned space mission scenarios including that of the International Space Station, the natural radiation environment encountered is a complex mixture of charged particles of galactic and solar origin and of those in the radiation belts trapped by the geomagnetic field (Stassinopoulos, 1988). In addition, secondary radiation is produced, such as proton recoils, neutrons, photons, and other by-products from the interaction of cosmic rays with the shielding material of the spacecraft and the astronaut's body (Wilson *et al.*, 1993).

In order to assess the risks posed by radiation to human health or other parts of the biosphere, estimates must be made of both the amount and type of radiation under consideration as well as the radiobiological effectiveness of the different components of the radiation. Ionizing radiation is measured in the S.I. unit of absorbed dose, the gray (Gy), with 1 Gy equal to the net absorption of 1 J in 1 kg of any material. Generally, the material absorbing the energy is assumed to be water, a material close in radiation absorption properties to those of tissue. The S.I. unit for the dose of non-ionizing radiation e.g., UV radiation, is J/m^2.

In manned space missions, the overall absorbed dose has been recorded at different locations both inside and outside of the spacecraft as well as at specific locations on the astronaut (Benton and Parnell, 1988; Reitz *et al.*, 1997). These measurements are complicated by many factors, including the diversity of the radiation types, the changes in the intensity of radiation due to the orbital parameters, the time of the mission within the solar cycle, the changing position of the spacecraft in orbit, and the complex and frequently changing effective shielding due to the movement of the astronauts within the spacecraft. A further complication in space radiation dosimetry lies in the fact that different types of radiation cause different amounts of biological damage per unit of absorbed dose. Therefore, in order to assess radiation risks in space, a biological weighting of the different types of radiation is required. Furthermore, potential interactions of radiation and other space flight factors including microgravity have been reported (Horneck, 1992). In order to assess the impact of space radiation on human health, biological dosimeters should be considered to be used in parallel with physical dosimeters because they directly weight the different components of radiation according to their biological effectiveness and potential interactions with other space flight factors. Their potential and advantages for such usage will be discussed in this paper.

Recognition of the hazardous nature of environmental UV-B radiation and the potential for future increased UV-B fluxes as a result of stratospheric ozone depletion has led to international actions (UNEP, 1991) and the establishment of national and international UV monitoring networks (WMO, 1994). Spectroradiometry is a common radiometric technique in UV-B monitoring. High demands on instrument specifications, such as high accuracy (especially at the edge of the solar spectrum in the UV-B range), high stray light suppression, high reproducibility and temperature stability are required. Frequent calibration with standard lamps and field intercomparisons with other spectroradiometers are indispensable (SCOPE, 1992; McKenzie *et al.*, 1993; Webb *et al.*, 1994). Because biological effects of UV radiation are strongly dependent on wavelength, spectral weighting functions must be applied to relate the physical doses to meaningful biologically weighted doses. This quantification of the biological effectiveness of solar radiation is complicated by several factors. Although ozone depletion affects only the UV-B edge of sunlight, depending on the tail of the action spectrum, the responses to UV-A and even to visible light might be important. Furthermore, the method assumes simple additivity of the various wavelength components by incorporating a multiplicative constant appropriate for each wavelength. This is based on the fact that most action spectra have been developed with monochromatic radiation. If interactions occur - and they have been reported for various biological effects (Jagger, 1985) - simple additivity is an inadequate basis for biologically weighted dosimetry and will not be a suitable indicator for the biological responses to the solar radiation reaching the surface of the Earth. In this report, the potential of biological dosimeters will be discussed that (i) directly weight the incident UV components of sunlight in relation to the effectiveness of different wavelengths and (ii) take into account any interactions between them.

If applied in parallel with physical dosimetry, biological dosimetry systems can provide valuable information because of the following reasons: (i) their ability to weight the different components of environmental radiation according to their biological efficacy; (ii) their ability to give a record of the accumulated radiation exposure of individuals; (iii) their high specificity; and (iv) their high sensitivity.

BIOLOGICAL WEIGHTING FUNCTIONS

Quality Factor for Ionizing Radiation

In order to calculate the biological responses to radiations of high linear energy transfer (LET), such as those experienced in space, radiation protection procedures have introduced the quality factor Q. Q is the biological weighting function of ionizing radiation and is related to the LET of the radiation (ICRP 1991) (Figure 1). It has been obtained from averaging the results of different experiments with different types of ionizing radiation. For X rays and γ rays Q is equal to 1. For a given dose of high-LET radiation the dose equivalent, H, is the amount of low-LET radiation (e.g., γ or X rays) necessary to produce a biological effect equivalent to that produced by high-LET radiation:

$$H = QD \tag{1}$$

with H = the dose equivalent, Q = the quality factor and D = the absorbed dose. The S.I. unit for the dose equivalent is Sievert (Sv). For a mixed radiation field composed of ionizing radiations of different radiation qualities, i, (as encountered in space), the dose equivalent to a given tissue, H_t, is defined as

$$H_t = \sum N_{t,i} Q_i D_{t,i} \, (Sv) \tag{2}$$

with $D_{t,i}$ = absorbed dose, deposited in the tissue t by the radiation i (Gy), Q_i = radiation quality which is described as a function of LET, and $N_{t,i}$ = a special factor which accounts for specific exposure conditions (e.g. dose rate, fractionated exposure, microgravity) or special tissue properties.

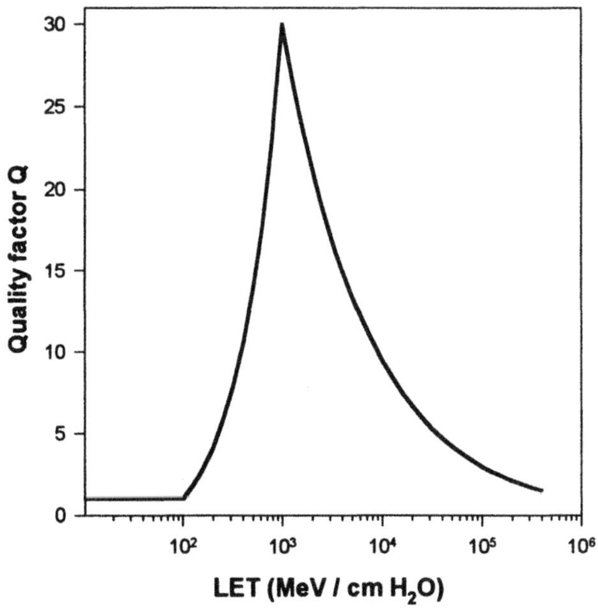

Fig. 1. Dependence of quality factor Q on LET (from ICRP 1991).

For manned missions in LEO, an average Q value of 2.9 has been assumed which takes into consideration the radiation field behind a 4 g/cm² Al shielding (Silberberg *et al.*, 1984; NCRP 1989). However, depending on the orbital parameters and the time of the mission during the solar cycle, the actual Q value, determined by Eq. 2 from the absorbed doses of the different components of the radiation field in space, can vary significantly (Bücker *et al.*, 1993; Reitz *et al.*, 1995).

Action Spectra for UV Radiation

To quantify the biological effectiveness of environmental UV radiation, the physical dose parameters have to be converted into biologically meaningful dose parameters. Biologically weighted radiometric quantities are derived from the spectral data by multiplying them with an action spectrum of a relevant photobiological reaction, e.g. DNA damage, erythema formation, skin cancer, reduced productivity of terrestrial plants, or UV sensitivity of aquatic ecosystems. The biologically effective irradiance E_{eff} (W_{eff}/m^2) is then determined as follows (Setlow, 1974):

$$E_{eff} = \int E_{\lambda}(\lambda) \cdot S_{\lambda}(\lambda) d\lambda \quad (3)$$

with $E_{\lambda}(\lambda)$ = solar spectral irradiance (W/m^2 nm), $S_{\lambda}(\lambda)$ = action spectrum (relative units), and λ = wavelength (nm). Integration of the biologically effective irradiance E_{eff} over time (e.g., a full day) gives the biologically effective dose H_{eff} (J_{eff}/m^2) (e.g., daily dose) (Horneck 1995). Figure 2 shows the action spectrum for the minimal erythema, which is recommended as reference action spectrum by The Commission International de l'Eclairage (CIE). It has been obtained by averaging over the spectral responses of various individuals of different skin types (McKinley and Diffey, 1987).

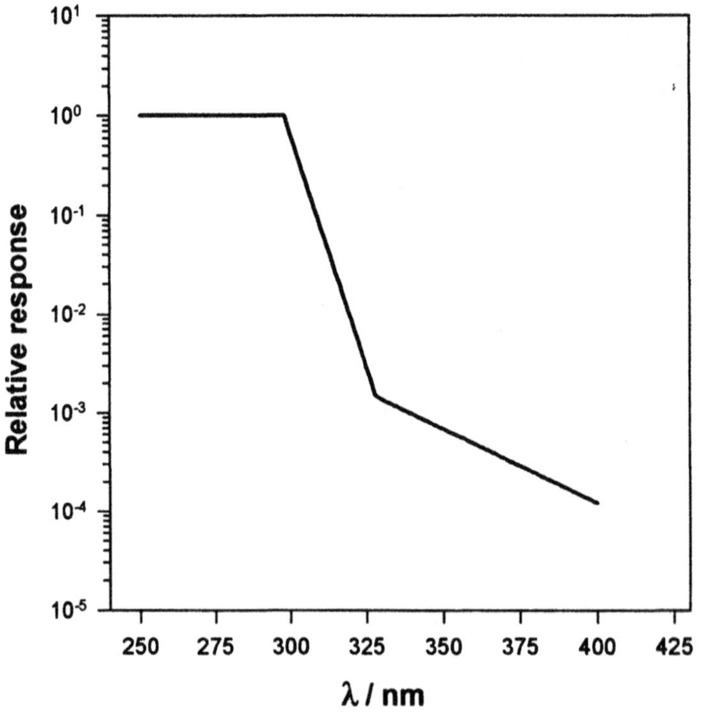

Fig. 2. Action spectrum for the minimal erythema which has been recommended by CIE as reference action spectrum for assessing the health impact of environmental UV radiation (data from McKinley and Diffey, 1987). Reference wavelength is λ = 298 nm.

BIOLOGICAL RADIATION MONITORING SYSTEMS

Intrinsic Biomarkers

Intrinsic biomarkers give a record of the individual radiation exposure (e.g., of humans, animals, plants or ecosystems) in measurable units. For monitoring personal exposure to ionizing radiation, the following intrinsic biomarkers for genetic or metabolic changes are already available: (i) chromosomal aberrations in human lymphocytes (Evans *et al.*, 1979; IAEA, 1986; Obe *et al.*, 1997; Straume and Bender, 1997); (ii) germ line minisatellite mutation rates or radiation induced apoptosis (Menz *et al.*, 1997); (iii) metabolic changes in serum, plasma or urine (e.g., serum lipids, lipoproteins, ratio of HDL/LDL cholesterol, lipoprotein lipase activity, lipid peroxides, melatonin, or antibody titers) (Ahlers, 1994; Ahlers *et al.*, 1995); (iv) hair follicle changes and decrease in hair thickness (Potten, 1993); (v) triacylglycerol-concentration in bone marrow (Ahlers, 1994); and (vi) glycogen concentration in liver (Ahlers *et al.*, 1995). Whereas the first four systems mentioned are non-invasive or require only blood samples for analysis, the latter two systems are invasive and therefore appropriate for radiation monitoring in animals only. Dose response relationships have been described for most of the intrinsic dosimetry systems.

In astronauts, after long-term space flights, an elevation of the frequencies of chromosomal aberrations in peripheral lymphocytes has been reported (Testard *et al.*, 1996; Obe *et al.*, 1997; Yang *et al.*, 1997). Obe *et al.* (1997) investigated lymphocytes of 7 astronauts that had spent several months on board the MIR space station. They showed that the frequency of dicentric chromosomes increased by a factor of approximately 3.5 compared to preflight control and that the observed frequencies agreed quite well with the expected values based on the absorbed doses and particle fluxes encountered by individual astronauts during the mission. These data suggest the feasibility of using chromosomal aberrations as a biological dosimeter for monitoring radiation exposure of astronauts. In these studies, the analysis of chromosomal aberrations was performed at the first mitosis following *in vitro* growth stimulation. Recently, premature chromosome condensation (PCC) technique has been developed that allows interphase chromosome painting (Durante, 1999) and the detection of non-rejoining chromatin breaks detection (Suzuki *et al.*, 1999) without going through the first mitosis. This method is especially relevant for biological dosimetry of astronauts that are exposed to high-LET heavy ions of cosmic radiation which induce interphase death and cell cycle delay.

For monitoring personal exposure to environmental UV radiation, the following intrinsic biomarkers may be applicable: (i) photoproducts induced in the skin (e.g., cyclobutadipyrimidines, (6-4) pyrimidine-pyrimidone adducts and their Dewar valence isomer) (Freeman, 1989; Chadwick *et al.*, 1995; Young *et al.*, 1997) and their repair rates; (ii) genes mutations in the skin (e.g., *p53* mutations that occur predominantly associated with skin cancer) (Daya-Grosjean *et al.*, 1995; Ananthaswamy *et al.*,1998); (iii) second messengers in skin cells (e.g., regulation of matrix metalloproteinases) (Brenneisen *et al.*, 1996); (iv) antibody titers in the blood; and (v) lens turbidity that can be measured by fluorescence. The first three methods mentioned are invasive, require biopsy and are therefore not applicable for large scale screening of UV exposure of humans. However, they should be used to calibrate extrinsic biological dosimeters.

Extrinsic biological dosimeters

Extrinsic biological dosimeters are especially suited for long-term monitoring of global changes in environmental UV radiation and its biological implications. They are also useful in monitoring UV exposure of individuals during out-door activities (e.g., skiing, hiking, gardening, or leisure) or those at special risk. Biological dosimeters automatically weight the incident UV components of sunlight relative to the biological effectiveness of the different wavelengths and any interactions between them. Ideally, the spectral response of the biological dosimeter is identical to that of the action spectrum of the photobiological effect under consideration. In this case, the biologically effective dose H_{eff} is equivalent to the incident dose of monochromatic radiation at a standard wavelength λ, which would produce the same response as the actual radiation under consideration (Tyrrell, 1980). It is given by the following relation:

$$H_{eff} = F \tag{4}$$

with H_{eff} = biologically effective dose (J_{eff}/m^2) and F = equivalent dose of monochromatic UV radiation producing the same biological response (J/m^2). So far, biological UV dosimeters of different levels of complexity are available, inlcuding (i) biomolecules (e.g., the uracile molecule or DNA) (Gróf *et al.*, 1996; Regan *et al.*, 1992; Yoshida and Regan, 1997); (ii) viruses (e.g. bacteriophage T1, T2, T4, or T7) (Rontó *et al.*, 1994); and (iii) bacteria (e.g. *E. coli*, spores of *Bacillus subtilis*, or *Euglena gracilis* phytoplankton) (Munakata, 1993; 1995; Horneck et al., 1996; reviewed in Horneck 1997). Their action spectra (Figure 3) agree quite well with that for DNA damage (Setlow, 1974). Since UV-induced cancer is probably initiated by photochemical changes of the DNA (Yarosh, 1992), a predominant mechanism of UV-B, these simple biological dosimeters are suitable to estimate the potential carcinogenic risk of an increased solar UV radiation.

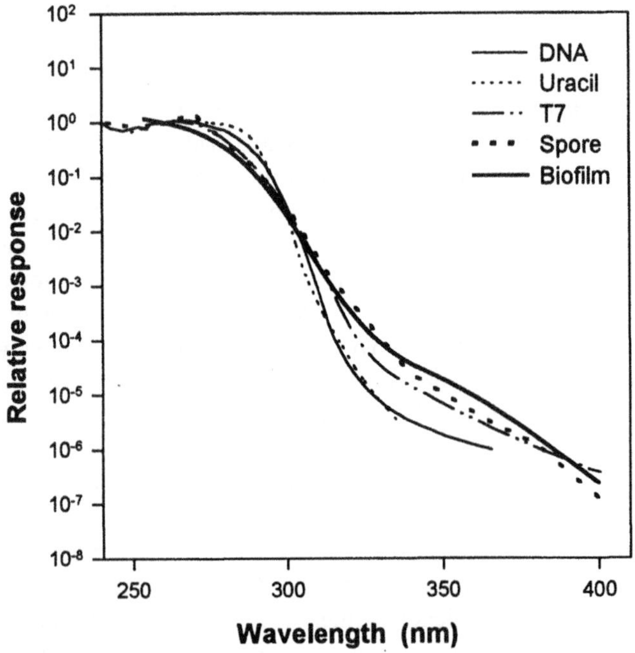

Fig. 3. Relative action spectra of different biological dosimeters (from Horneck, 1997)

The biological UV dosimeters are simple, robust and functional indicators of systems at risk (e.g., DNA damage, photosynthetic impairment, reduced biological activity, or loss in vitality). Most of them are well characterized concerning their photobiological (e.g., action spectra) and radiometric properties and have been cross-calibrated with other UV radiometers.

Biosensors and Bioassays for Genotoxicity

Increasing levels of environmental pollution demand specific and sensitive detection methods for environmental toxins. Cellular bioassays for genotoxic assessment are based on DNA damage induced in target cells. The results are frequently used to infer the mutagenic and carcinogenic hazard posed to humans and other organisms. The Ames assay (Ames, 1979) uses a set of auxotrophic strains of *Salmonella typhimu-*

rium that revert to histidine prototrophy upon exposure to mutagens of specific mechanisms. The assay has been validated in studies using several hundred chemicals (Mersch-Sundermann *et al.*, 1994).

Table 1. Comparison of the Sensitivity of Different Bioassays for Genotoxicity (from Horneck *et al.*, 1998).

Genotoxin	Lower limit of detection[a] (M) for chemicals, (Gy) for γ rays			Revertants (nM^{-1})[b] or lower limit of detection[a] (Gy)
	SOS *lux* test	SOS chromotest	*umu* test	Ames test
MMC	5.0×10^{-9}	1.7×10^{-8}	1.5×10^{-7}	1200
DMS	7.5×10^{-6}	6.7×10^{-6}	3.0×10^{-4}	0.1
MNNG	7.1×10^{-7}	5.0×10^{-7}	2.0×10^{-6}	44
γ rays	2.56	<5	n.d.	3 - 4

[a] Dose of the genotoxin to increase the response by a factor of 2 over background.
[b] Sensitivity to the chemicals MMC, DMS, and MNNG.
M = mole; MMC = mitomycin C; DMS = dimethylsulfate; MNNG = N-methyl-N'-nitro-N-nitrosoguanidine; n.d. = not determined

SOS-dependent bacterial test systems make use of the fact that in response to DNA-damaging agents, a cascade of functions including the transcription of more than 15 repair enzymes known as the SOS response is induced. This includes a number of proteins involved in mutagenesis such as RecA and UmuC/D (Witkin, 1976). In the SOS chromotest (Quillardet and Hofnung, 1993) *E. coli* PQ37 cells with the structural genes for β-galactosidase, *lacZ*, under the control of a SOS controlled gene, *sulA*, are used as test system for genotoxicity. The *umu* test makes use of a recombinant *S. typhimurium* TY 1535 (pSK1002) strain with a fused *umuC:lacZ* operon carried on the plasmid (Oda *et al.*, 1995). In both test systems, the SOS induction potency is determined from a colorimetric assay for β-galactosidase in response to a genotoxin. The SOS *lux* assay (Ptitsyn *et al.*, 1997) utilizes the SOS system as receptor which is sensitive to DNA damage and the bioluminescence system as rapid optical reporter. In Table 1 the sensitivity of different bioassays for genotoxicity is compared. These bioassays are potentially useful for *in situ* and continuous detection of genotoxins.

RECOMMENDED RESEARCH AND APPROACH

During the NATO Advanced Research Workshop on Fundamentals for the Assessment of Risks from Environmental Radiation, held at Brno, Czech Republic, from October 6th to 10th 1997, several recommendations for future research activities in the field of biological monitoring of radiation exposure were suggested as outlined in the following:

- For monitoring the personal or individual radiation exposure, efforts should be made towards the development of methods with, and applications of, *in situ* biomarkers.
- The responses of extrinsic biological dosimeters should be correlated with relevant human health processes and intrinsic biomarkers, e.g., *in situ* DNA lesion measurements in the skin, especially in case of personal dosimeters.

- Further biological dosimeters and biosensors for genotoxins should be designed that are based on well characterized mechanisms of response with well defined weighting of the different components of environmental radiation.
- The applicability of extrinsic biological dosimeters for the detection of ionizing radiation should be investigated.
- Biological dosimeters and indicators should be included in long-term monitoring of global changes in environmental UV radiation and its biological applications.
- Biological dosimeters should be included in monitoring UV exposure of individuals either at risk or during different out-door activities (e.g., skiing, hiking, gardening, leisure).

ACKNOWLEDGEMENT

The paper gives a résumé of the recommendations of a task group at the NATO Advanced Research Workshop on Fundamentals for the Assessment of Risks from Environmental Radiation held at Brno, Czech Republic, from October 6th to 10th, 1997. Members of the task group are I. Ahlers, L. O. Björn, D.-P. Häder, B. Kaina, O. V. Komova, S. Kozubek, G. Obe, C. S. Potten, L. R. Ptitsyn, H. Rink, M. Wlaschek and T. C. Yang. Their contributions are especially appreciated. Part of the study was supported by a grant of the European Commission (ENV4-CT95-0044).

REFERENCES

Ahlers, I., Changes in Whole-Body Metabolic Parameters Associated with Radiation, *Adv. Space Res.* **14**, (10)531-(10)539 (1994).

Ahlers, I., E. Ahlersova, M. Sabol, P. Solar, M. Kassayova, and B. Smajda, Metabolic and Hormonal Changes in Continuously Irradiated Rats are Influenced by Exogenous Melatonin, in *Radiation Biology and its Application in Space Research*, edited by S. Kozubek and G. Horneck, pp. 139-142, Kimaro, Brno, Czech Republic (1995).

Ames, B. N., Identifying Environmental Chemicals causing Mutations and Cancer, *Science*, **204**, 587-593 (1979).

Ananthaswamy, H. N., A. Fourtanier, R. L. Evans, S. Tison, C. Medaisko, S. E. Ullrich, and M. L. Kripke, *p53* Mutations in Hairless SKH-hr1 Mouse Skin Tumors Induced by a Solar Simulator, *Photochem. Photobiol.*, **67**, 227-232 (1998).

Benton, E. V., and T. A. Parnell, Space Radiation Dosimetry on U.S. and Soviet Manned Missions, in *Terrestrial Radiation and Its Biological Effects*, edited by P. D. McCormack, C. E. Swenberg, and H. Bücker pp. 729-794, Plenum, NewYork, NY, USA (1988).

Brenneisen, P., J. Oh, M. Wlaschek, J. Wenk, K. Briviba, C. Hommel, G. Hermann, H. Sies, and K. Scharffetter-Kochanek, Ultraviolet B Wavelength Dependence for the Regulation of Two Major Matrix-Metalloproteinases and their Inhibitor TIMP-1 in Human Dermal Fibroblasts, *Photochem. Photobiol.*, **64**, 649-657 (1996).

Bücker, H., G. Horneck, R. Facius, and G. Reitz, Radiation Exposure in Manned Space Flight, *Kerntechnik*, **58**, 229-234 (1993).

Chadwick, C. A., C. S. Potten, O. Nikaido, T. Matsunaga, C. Proby, and A. R. Young, The Detection of Cyclobutane Thymine Dimers, (6-4) Photolesions and the Dewar Photoisomers in Sections of UV-Irradiated Human Skin Using Specific Antibodies, and the Demonstration of the Depth Penetration Effect, *J. Photochem. Photobiol. B: Biol.*, **28**, 163-179 (1995).

Daya-Grosjean, L., N. Dumaz, and A. Sarasin, The Specificity of *p53* Mutation Spectra in Sunlight Induced Human Cancers, *J. Photochem. Photobiol. B: Biol.*, **28**, 115-124 (1995).

Durante, M., Biodosimetry of Heavy ions by Interphase Chromosome Painting, *Adv. Space Res.* (this issue) (1999).

Evans, H. J., K. E. Buckton, G. E. Hamilton, and A. Carothers, Radiation-Induced Chromosome Aberrations in Nuclear-Dockyard Workers, *Nature*, **277**, 531-534 (1979).

Freeman, S. E., H. Hacham, R. W. Gange, D. J. Maytum, J. C. Sutherland, and B. M. Sutherland, Wavelength Dependence of Pyrimidine Dimer Formation in DNA of Human Skin Irradiated *in situ* with Ultraviolet Light, *Proc. Natl. Acad. Sci. USA*, **86**, 5605-5609 (1989).

Gróf, P., S. Gáspár, and Gy. Rontó, Use of Uracil Thin Layer for Measuring Biologically Effective UV Dose, *Photochem. Photobiol.* **64**, 800-806 (1996).

Horneck, G., Radiobiological Experiments in Space: a Review. *Nucl. Tracks Radiat. Meas.*, **20**, 185-205 (1992).

Horneck, G., Quantification of the Biological Effectiveness of Environmental UV Radiation, *J. Photochem. Photobiol. B:Biol.*, **31**, 43-49 (1995).

Horneck, G., P. Rettberg, E. Rabbow, W. Strauch, G. Seckmeyer, R. Facius, G. Reitz, K. Strauch, and J.U. Schott, Biological Dosimetry of Solar Radiation for Different Simulated Ozone Column Thickness. *J. Photochem. Photobiol. B:Biol.*, **32**, 189-196 (1996).

Horneck, G., Biological UV Dosimetry, in *The Effects of Ozone Depletion on Aquatic Ecosystems*, edited by D.P. Häder, pp. 119-142, R.G. Landes Company, Austin, TX, USA (1997).

Horneck, G., L. R. Ptitsyn, P. Rettberg, O. Komova, S. Kozubek, and E. A. Krasavin, Recombinant *Escherichia coli* Cells as Biodetector System for Genotoxicity, in *Biosensors for Environmental Diagnostics*, edited by B. Hock, D. Barceló, K. Camman, P.-D. Hansen, and A.P.F. Turner, pp. 215-232, B.G. Teubner, Stuttgart, Germany (1998).

IAEA, International Atomic Energy Agency, *Biological Dosimetry: Chromosomal Aberration Analysis for Dose Assessment*, Technical Reports Series No. 260, IAEA, Vienna, Austria (1986).

ICRP, *Recommendations of the International Commission on Radiological Protection*, Report 60, Pergamon Press, Oxford, UK (1991).

Jagger, J., *Solar Actions on Living Cells*, Praeger, New York, NY, USA (1985)

McKenzie, R. L., M. Kotkamp, G. Seckmeyer, R. Erb, C. R. Roy, H. P. Gies, and S. J. Toomey, First Southern Hemisphere Intercomparison of Measured Solar UV Spectra, *Geophys. Res. Lett.*, **20**, 2223-2226 (1993).

McKinley, A. F., and B. L. Diffey, A Reference Action Spectrum for Ultraviolet Induced Erythema in Human Skin, *CIE J.*, **6**, 17-22 (1987).

Menz, R., R. Andres, B. Larsson, M. Ozsahin, K. Trott, and N.E.A. Crompton, Biological Dosimetry: the Potential of Radiation-Induced Apoptosis in Human T-Lymphocytes, *Radiat. Environm. Biophys.*, **36**, 175-181 (1997).

Mersch-Sundermann, V., U. Schneider, G. Klopman, and H. S. Rosenkranz, SOS Induction in *Escherichia coli* and *Salmonella* Mutagenicity: a Comparison Using 330 Compounds, *Mutagenesis*, **9**, 205-224 (1994).

Munakata, N., Biologically Effective Dose of Solar Ultraviolet Radiation Estimated by Spore Dosimetry in Tokyo since 1980, *Photochem. Photobiol.*, **58**, 386-392 (1993).

Munakata, N., Continual Increase in Biologically Effective Dose of Solar UV Radiation Determined by Spore Dosimetry from 1980 to 1993 in Tokyo, *J. Photochem. Photobiol. B: Biol.*, **31**, 63-68 (1995).

NCRP, National Council on Radiation Protection and Measurements, *Guidance on Radiation Received in Space Activities*, NCRP Report 98, NCRP, Bethesda, Maryland, USA (1989)

Obe, G., I. Johannes, C. Johannes, K. Hallmann, G.Reitz, and R. Facius, Chromosomal Aberrations in Blood Lymphocytes of Astronauts after Long-Term Space Flights, *Int. J. Radiat. Biol.*, **72**, 726-734 (1997).

Oda, Y., H. Yamazaki, M. Watanabe, T. Nohmi, and T. Shimada, Development of Highly Sensitive *umu* Test System: Rapid Detection of Genotoxicity of Promutagenic Aromatic Amines by *Salmonella thyphimurium* Strain NM2009 Possessing High O-Acetyltransferase Activity, *Mutat. Res.*, **334**, 145-156 (1995).

Potten, C. S., Hair Cortical Cell Counts (HCCC), a New Sensitive *in vivo* Assay with Possible Applications for Biological Dosimetry, *Int. J. Radiat. Biol.*, **63**, 91-95 (1993).

Ptitsyn L. R., G. Horneck, O. Komova, S. Kozubek, E. A. Krasavin, M. Bonev, and P. Rettberg, A Biosensor for Environmental Genotoxin Screening Based on an SOS *lux* Assay in Recombinant *Escherichia coli* Cells, *Appl. Environm. Microbiol.*, **63**, 4377-4384 (1997).

Quillardet, P., and M. Hofnung, The SOS Chromotest: a Review, *Mutat. Res.* **297**, 235-279 (1993)

Regan, J. D., W. L. Carrier, H. Gucinski, B. L. Olla, H. Yoshida, R. K. Fujimura, and R.I. Wicklund, DNA as a Solar Dosimeter in the Ocean, *Photochem. Photobiol.*, **56**, 35-42 (1992).

Reitz, G., R. Beaujean, and M. Leicher, Dosimetric Measurements in Manned Missions, Acta Astronaut., **36**, 517-526 (1995).

Reitz, G., R. Beaujean, J. Kopp, M. Leicher. K. Strauch, and C. Heilmann, Results of Dosimetric Measurements in Space Missions, *Radiat. Protect. Dosim.*, **70**, 413-418 (1997).

Rontó, G., S. Gáspár, P. Gróf, A. Bérces, and Z. Gugolya, Ultraviolet Dosimetry in Outdoor Measurements Based on Bacteriophage T7 as a Biosensor, *Photochem. Photobiol.*, **59**, 209-214 (1994).

Rontó, G, P. Gróf, and S. Gáspár, Biological UV Dosimetry - a Comprehensive Problem, *J. Photochem. Photobiol. B:Biol.*, **31**, 51-56 (1995).

Rowland, F. S., Chlorofluorocarbons and the Depletion of Stratospheric Ozone, *Am. Scientist*, **77**, 36-46 (1989).

SCOPE Report, *Effects of Increased Ultraviolet Radiation on Biological Systems*, SCOPE Secretariat, Paris, France (1992).

Setlow, R. B., The Wavelengths in Sunlight Effective in Producing Skin Cancer: a Theoretical Analysis, *Proc. Natl. Acad. Sci. USA*, **71**, 3363-3366 (1974).

Silberberg, R., C.H. Tsao, J.H.Jr. Adams, and J.R. Letaw, LET-Distributions and Doses of HZE Radiation Components in Near-Earth Orbits, *Adv. Space Res.*, **4**(10), 143-151 (1984).

Stassinopoulos, E. G., The Earth's Trapped and Transient Space Radiation, in *Terrestrial Space Radiation and Its Biological Effects*, edited by P. D. McCormack, C. E. Swenberg and H. Bücker, pp. 5-35, Plenum Press, New York, NY, USA (1988).

Straume T., and M. A. Bender, Issues in Cytogenetic Biological Dosimetry: Emphasis on Radiation Environments in Space, *Radiat. Res.* **148**, 560-570 (1997).

Suzuki, M., Y. Kase, T. Nakano, T. Kanai, and K. Ando, The Induction of Non-Rejoining PCC Breaks as Biodosimetry for Cell Killing by Carbon Ions, *Adv. Sapce Res.* (this issue) (1999).

Testard, I., M. Ricoul, F. Hoffschir, A. Flury-Herard, B. Dutrillaux, B. Federenko, V. Gerasimenko, and L. Sabatier, Radiation-Induced Chromosome Damage in Astronauts' Lymphocytes, *Int. J. Radiat. Biol.*, **70**, 403-411 (1996).

Tyrrell, R.M., Solar Dosimetry and Weighting Factors, *Photochem. Photobiol.*, **31**, 421-422 (1980).

UNEP, International Committee on Effects of Ozone Depletion, *UNEP Report on the Environmental Effects of Ozone Depletion*, UNEP, New York, NY, USA (1991).

Webb, A. R., B. G. Gardiner, M. Blumthaler, P. Forster, M. Huber, and P. J. Kirsch, A Laboratory Investigation of Two Ultraviolet Spectroradiometers, *Photochem. Photobiol.*, **69**, 84-90 (1994).

Wilson, J. W., L. W. Townsend, W. Schimmerling, G. S. Khandelwal, F. Khan, J. E. Nealy, F. A. Cucinotta, and J. W. Norbury, Transport Methods and Interactions for Space Radiations, in *Biological Effects and Physics of Solar and Galactic Cosmic Radiation, Part B*, edited by C. E. Swenberg, G. Horneck and E.G. Stassinopoulos, pp. 187-786, Plenum Press, New York, NY, USA (1993).

Witkin, E.M., Ultraviolet Mutagenesis and Inducible DNA Repair in *Escherichia Coli, Bacteriol. Rev*, **40**, 869-907 (1976).

WMO, World Meteorological Organization, Global Atmosphere Watch, *Report of the WMO Meeting of Experts on UV-B Measurements, Data Quality and Standardization of UV Indices*, Report No. 95, WMO/TD-NO. 625, Geneva, Switzerland (1994).

Yang, T.C., K. George, A.S. Johnson, M. Durante, and B.S. Fedorenko, Biodosimetry Results from Space Flight Mir-18, *Rad. Res.*, **148**, S17-S23 (1997).

Yarosh, D. B., The Role of DNA Damage and UV-Induced Cytokines in Skin Cancer, *J. Photochem. Photobiol. B: Biol.*, **16**, 91-94 (1992).

Yoshida, H., and J. D. Regan, UVB DNA Dosimeters Analyzed by Polymerase Chain Reactions, *Photochem. Photobiol.*, **66**, 82-88 (1997).

Young A. R., C. A. Chadwick, G. I. Harrison, J. Ramsden, and C. S. Potten, Thymine Dimer Action Spectra in Different Human Epidermal Layers and their Relationship with Erythema Action Spectra, *Photochem Photobiol.*, **65**, 80S (1997).

Pergamon

Adv. Space Res. Vol. 22, No. 12, pp. 1643–1652, 1998

Printed in Great Britain
0273-1177/98 $19.00 + 0.00

PII: S0273-1177(99)00029-0

BIOLOGICAL DOSIMETRY TO DETERMINE THE UV RADIATION CLIMATE INSIDE THE MIR STATION AND ITS ROLE IN VITAMIN D BIOSYNTHESIS

P. Rettberg[1], G. Horneck[1], A. Zittermann[2], M. Heer[1]

[1]DLR, Institut für Luft- und Raumfahrtmedizin, Strahlenbiologie, D - 51170 Köln, Germany
[2]Institut für Ernährungswissenschaft, Universität Bonn, Germany

ABSTRACT

The vitamin D synthesis in the human skin, is absolutely dependent on UVB radiation. Natural UVB from sunlight is normally absent in the closed environment of a space station like MIR. Therefore it was necessary to investigate the UV radiation climate inside the station resulting from different lamps as well as from occasional solar irradiation behind a UV-transparent quartz window. Biofilms, biologically weighting and integrating UV dosimeters successfully applied on Earth (e.g. in Antarctica) and in space (D-2, Biopan I) were used to determine the biological effectiveness of the UV radiation climate at different locations in the space station. Biofilms were also used to determine the personal UV dose of an individual cosmonaut. These UV data were correlated with the concentration of vitamin D in the cosmonaut's blood and the dietary vitamin D intake. The results showed that the UV radiation climate inside the Mir station is not sufficient for an adequate supply of vitamin D, which should therefore be secured either by vitamin D supplementat and/or by the regular exposure to special UV lamps like those in sun-beds. The use of natural solar UV radiation through the quartz window for 'sunbathing' is dangerous and should be avoided even for short exposure periods.

INTRODUCTION

Photobiological reactions induced by ultraviolet (UV) radiation can be found in every biological system which is exposed to UV. The effects of solar UV radiation are manifold and concern individual organisms as well as whole ecosystems. Most of the health effects on humans are harmful. Exposure to the sun is known to be associated with various skin cancers, accelerated skin aging, cataract formation and other eye diseases. The UV induced suppression of the immune system may affect an individual's ability to resist infectious diseases as well as influence the effectiveness of vaccinations (WHO, 1994).

However, UV radiation has also beneficial effects. One of these is the cutaneous synthesis of vitamin D which is actually a steroid hormone and not a vitamin. The synthesis of the biologically active metabolites of vitamin D_3, the chemical species of vitamin D occurring in mammals, requires several steps of chemical reactions which take place in different organs (see Figure 1). The first step, the photochemical conversion of 7-dehydrocholesterol (7DHC) to pre-vitamin D_3 in the epidermis, absolutely requires UV radiation of wavelengths shorter than 315 nm, that is UVB and the longer wavelengths of UVC (Webb, 1993). This is

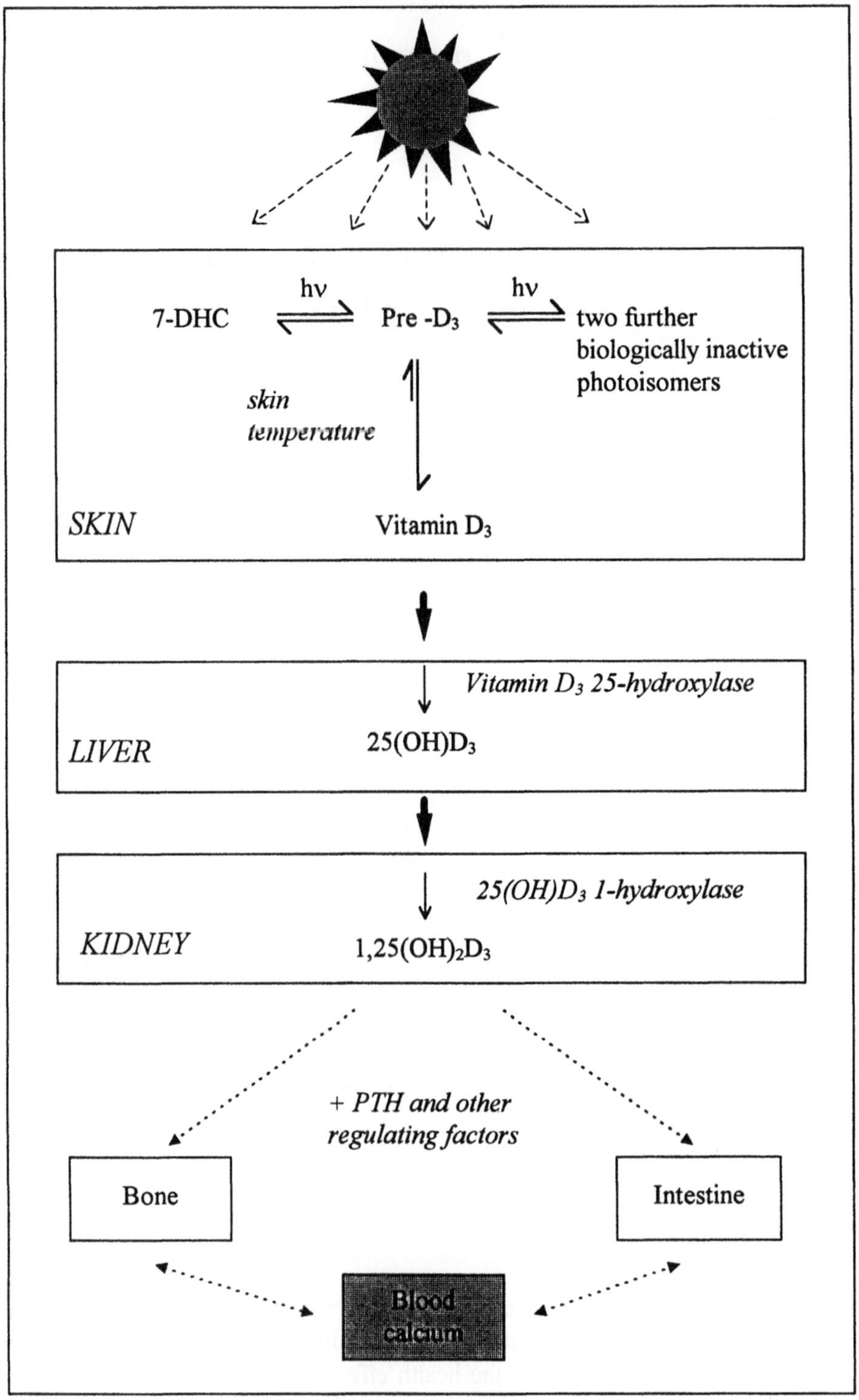

***Fig. 1.** Simplified diagram of vitamin D_3 formation and metabolic pathway to its biologically active metabolites*

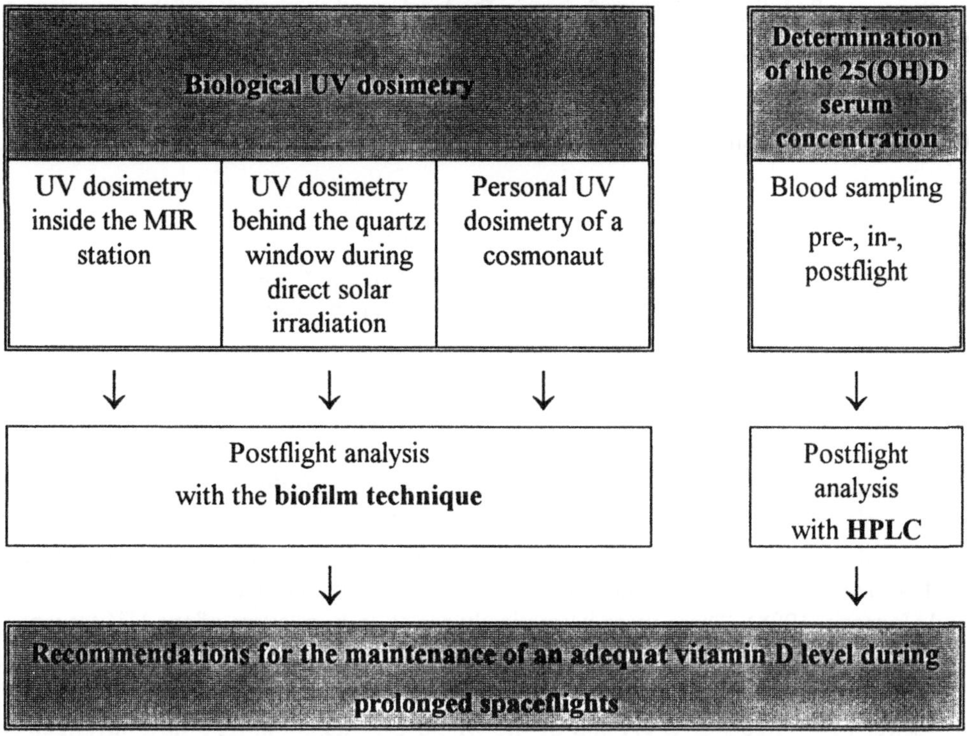

Fig. 2. Schematic overview of the experiment UVE

the same wavelength range which also causes the deleterious effects on humans. After a thermal conversion reaction in the skin to vitamin D_3 it is transported through the blood by the vitamin D-binding protein (DBP) to the liver where a first hydroxylation reaction to 25-hydroxy vitamin D_3 ($25(OH)D_3$) takes place. A second one is performed in the kidneys and results in the metabolically active form 1,25-dihydroxy vitamin D_3 ($1,25(OH)_2D_3$). This compound acts as a classical steroid hormone by binding specifically to a nuclear receptor in the target cells, thereby inducing or repressing several genes. In addition to the photochemical synthesis of pre-vitamin D_3 in the skin vitamin D can also be absorbed from food in the intestine, but in the normal human population the photochemical cutaneous synthesis is the main source of vitamin D.

Vitamin D and its active metabolites have many different functions in the human body. The most important one is the regulation of mineral homeostasis in target tissues intestine, bone and kidney, and especially the control of the calcium and phosphorus serum levels (Maiyar and Norman, 1991, for review). Changes in calcium metabolism have a strong influence on bone metabolism which is changed in microgravity during spaceflights. A significant loss of bone mass can be observed even after short-term space missions (Organov *et al.*, 1992, Caillot-Augusseau *et al.*, 1998). The influence of vitamin D and of possible changes in its level on bone metabolism during prolonged spaceflights are largely unknown.

During the German-Russian MIR'97 mission the UV radiation climate inside the Russian space station Mir was investigated in the UVE Experiment to determine whether UV radiation onboard Mir is sufficient for a cutaneous synthesis of vitamin D. The sources of UV radiation onboard may be the sun shining through a UV-transparent quartz window or artificial UV lamps. The biological UV dosimetry (Horneck, 1995) with the DLR-biofilm (Quintern *et al.*, 1992) was done at different locations in the space station and performed as personal UV dosimetry (Rettberg and Horneck, 1998) for one of the cosmonauts. In addition the serum levels of 25(OH)D of the same cosmonaut were determined before, during, and after the mission and compared to the controlled dietary intake of vitamin D. A schematic overview of the experiment is shown in Figure 2.

EXPERIMENTAL PROCEDURE

MIR'97 Mission

The MIR'97 mission with a German cosmonaut was launched on February 10, 1997, 15:09h MET in Baikonur, Kasachstan, with Sojus-TM25, and ended on March 2, 1997, 7:50h MET in Dschegasgan, Kasachstan. The duration was 19d 16h 41min.

Biological UV dosimetry

Method: The biological UV dosimeter 'biofilm' consists of *B. subtilis* spores as UV-sensitive target immobilized on a plastic sheet (Quintern *et al., 1992,* Patent No. DE 40 39 002 A1). The wavelength-dependent sensitivity of the biofilm, the action spectrum, is comparable to the action spectrum of inactivation of human keratinocytes. In the experiment UVE biofilms were used in so called biofilm stacks (Rettberg and Horneck, 1998). These were hermetically sealed in two layers of the UV-transparent plastic foil Whirlpak™. The biofilm stacks contained the biofilm itself and different layers of neutral density filters (Nytrel™ 20 and 40) to extend the dynamic range of the dosimeter and/or a cut-off filter (Mylar™) to discriminate between the effects of UVA and of the whole UV radiation. The filter arrangement and the resulting averaged biologically weighted transmission of the entrance optics above the surface of the biofilm are shown in Figure 3. Biofilms were transported, stored and exposed at room temperature. The exposure was performed in small plastic housings (52.5 x 41.5 x 7.5 mm^3, see Figure 4, Patent No. DE 195 35 273). Nine biofilm housings and eleven biofilm stacks were transported to the MIR station. After exposure the biofilms were brought back to the laboratory. Unexposed areas on each biofilm were

2 x Whirlpak™ NYTREL 20™ NYTREL 40™ Mylar™	2 x Whirlpak™ NYTREL 20™ NYTREL 40™
2 x Whirlpak™ NYTREL 40™ Mylar™	2 x Whirlpak™ NYTREL 40™
2 x Whirlpak™ Mylar™	2 x Whirlpak™

⇔

16 % T	20 % T
21 % T	27 % T
35 % T	44 % T

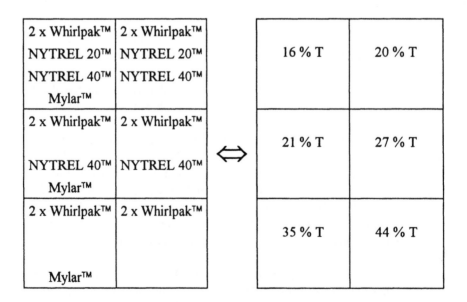

Fig. 3. Filter combinations and resulting transmissions in the biofilm stacks

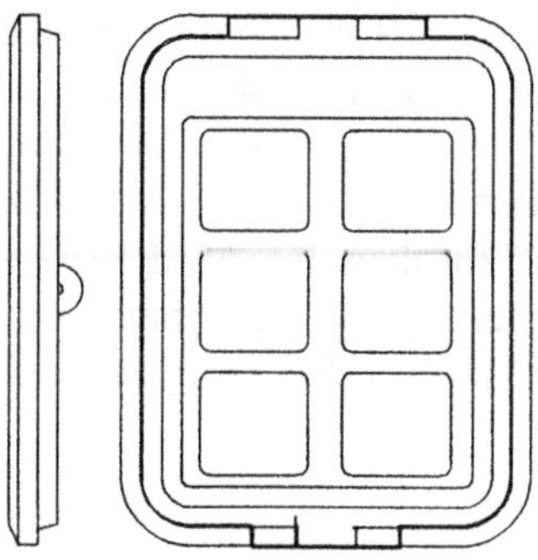

Fig. 4. Plastic housing of the biofilm stack

used for calibration with a standard UV source. The biofilms were then developed in nutrient medium under standardized conditions as described (Quintern et al., 1992). During the period of development the bacterial spores which were not or only slightly damaged by the previous UV exposure were able to germinate and and the outgrowing bacteria could multiply inside the biofilm. After fixation and staining of the biomass formed inside the biofilm, the quantitative analysis of the optical density of the measurement and calibration areas on each biofilm was performed with an image analysis system using a CCD camera and a data calculation unit to generate the biologically effective UV doses. With the filter combination used in this experiment, the lower limit of detection is about 20 $J_{eff}\ m^{-2}$, the upper limit of detection is about 500 $J_{eff}\ m^{-2}$.

Experiment Design and Performance: For the experiment UVE one biofilm dosimeter was used as personal UV dosimeter for the German cosmonaut from flight day 5 through 19 (15 d). Six biofilm dosimeters (3 x 2 in parallel) were used as stationary UV dosimeters at different locations in the MIR station in the base module from flight day 5 to 19 (14 d) fixed on panel 117, 433 and 329. Behind the quartz window in the base modul (illuminator 9) on flight day 5 two biofilm dosimeters were exposed in parallel during direct solar irradiation for 30 s, then the biofilm stacks in this two housing were exchanged and two other biofilm stacks were exposed for 120 s.

Determination of Serum Levels of 25(OH)D

Nutrient intake: During the baseline data collections (BDC's) and inflight, the German cosmonaut was on a constant energy and nutrient intake. In addition, he received a daily supplement of 16.6 µg of vitamin D_2 contained in two tablets. The application of vitamin D_2, which is not synthesized in humans but has the same biological activity as vitamin D_3 from cutaneous synthesis, allows one to differentiate between the vitamin D metabolites resulting from either cutaneous synthesis induced by UV or from dietary intake. The vitamin D_2 supplement began from preflight BDC III in June/July 1996 stopped after reentry.

Method: For determination of 25-hydroxyvitamin D metabolites ($25(OH)D_2$ and $25(OH)D_3$), serum samples were prepared by adding one volume of acetonitril, followed by an extraction and a purification step with C_{18} cartridges (Reinhardt et al., 1984). The $25(OH)D_2$ and $25(OH)D_3$ metabolites were then separated by high performance liquid chromatography using a silica column with hexan/isopropanol (98/2) as solvent. All samples were measured in duplicate during the same assay sequence. Standard solutions of $25(OH)D_2$ and $25(OH)D_3$ were handled in the same way. The vitamin D metabolites were quantified using a tritium labeled protein binding assay with rat serum as receptor (Hollis,1984). The intra- and interassay coefficients of variation (CV) were 2.3% and 7.5%, respectively.

Experiment Design and Performance: Fasting blood sampling of the German cosmonaut was scheduled during a preflight basic data collection (BDC) in June/July 1996 (BDC III) on days L-229, L-227, L-225,

L-221 and L-219 (229, 227, 225, 221 and 219 days before the launch of the space flight) and during a preflight BDC in December 1996 (BDC VI) on L-66. During the mission fasting blood samples were obtained on days 4, 8, 10, 13, 16 and 18. After reentry, an early control specimen was drawn on R+12 (12 days after the return of the space flight) during BDC XI. Additionally, a sequence of data collection was performed in August 1997 (BDC XIV) on days R+165, R+167, R+169, R+171 and R+173. All blood specimen were collected into serum monovettes, centrifuged for 30 min, aliquoted consecutively and frozen at -20°C (during the flight) and at -80°C (after return) until assay.

RESULTS

Biological UV dosimetry

The results of the biological UV dosimetry are given in Table 1. In no case the biologically effective UVA dose exceeded the lower limit of detection of 20 $J_{eff}\ m^{-2}$. In contrast, the dose of total UV varied significantly, depending on the position of the dosimeter. The highest UV doses were measured behind the quartz window (illuminator 9) during a period of direct insolation. Exposure for a 30 s period resulted in a biologically effective dose of 180 ± 7 $J_{eff}\ m^{-2}$ (average of two biofilms D-UVE-008 and -009). After an exposure of 120 s the total UV dose exceeded the upper limit of detection of the biofilm dosimeter of 500 $J_{eff}\ m^{-2}$. In contrast, almost all biofilm dosimeters exposed in the living area (panel 117, 329 and 433) during the whole mission showed a UV radiation dose below the limit of detection. Only dosimeter D-UVE-005 exposed in parallel to D-UVE-004 indicated a slightly higher UV dose of 22.6 ± 9.1 $J_{eff}\ m^{-2}$. However, the averaged dose of both parallel dosimeters resided below the lower limit of detection. Hence, after 15 days, the biologically effective UV radiation at any site in the Mir station except behind the quartz window is neglectible. It is interesting to note that the total UV dose of the cosmonaut for the whole mission was 30.0 ± 6.3 $J_{eff}\ m^{-2}$.

Table 1. Biologically effective UV doses

UV dosimeter	location	irradiation period	UVA / $J_{eff}\ m^{-2}$	total UV / $J_{eff}\ m^{-2}$
D-UVE-001	personal dosimeter	15 days	<20	30.0 ± 6.3
D-UVE-002	panel 433	15 days	<20	<20
D-UVE-003	panel 433	15 days	<20	<20
D-UVE-004	panel 117	15 days	<20	<20
D-UVE-005	panel 117	15 days	<20	22.6 ± 9.1
D-UVE-006	panel 329	15 days	<20	<20
D-UVE-007	panel 329	15 days	<20	<20
D-UVE-008	illuminator 9	30 s	<20	187 ± 7
D-UVE-009	illuminator 9	30 s	<20	172 ± 7
D-UVE-010	illuminator 9	120 s	<20	>500
D-UVE-011	illuminator 9	120 s	<20	>500

Serum Levels of 25(OH)D

Levels of the pre-, in- and postflight $25(OH)D_2$ and $25(OH)D_3$ are shown in Figure 5. The serum $25(OH)D_2$ concentrations remained virtually constant at 3.3 ± 0.3 ng ml^{-1} during preflight BDC III and showed a marked increase during BDC VI, inflight and during BDC XI. $25(OH)D_2$ levels of postflight BDC XIV were comparable with preflight BDC III with a slight increase occurring at end of postflight BDC XIV, six months after the mission. Mean inflight $25(OH)D_3$ concentrations were reduced in comparison to mean postflight values of BDC XIV, while no significant changes in total 25(OH)D ($25(OH)D_2$ and $25(OH)D_3$) levels did occur.

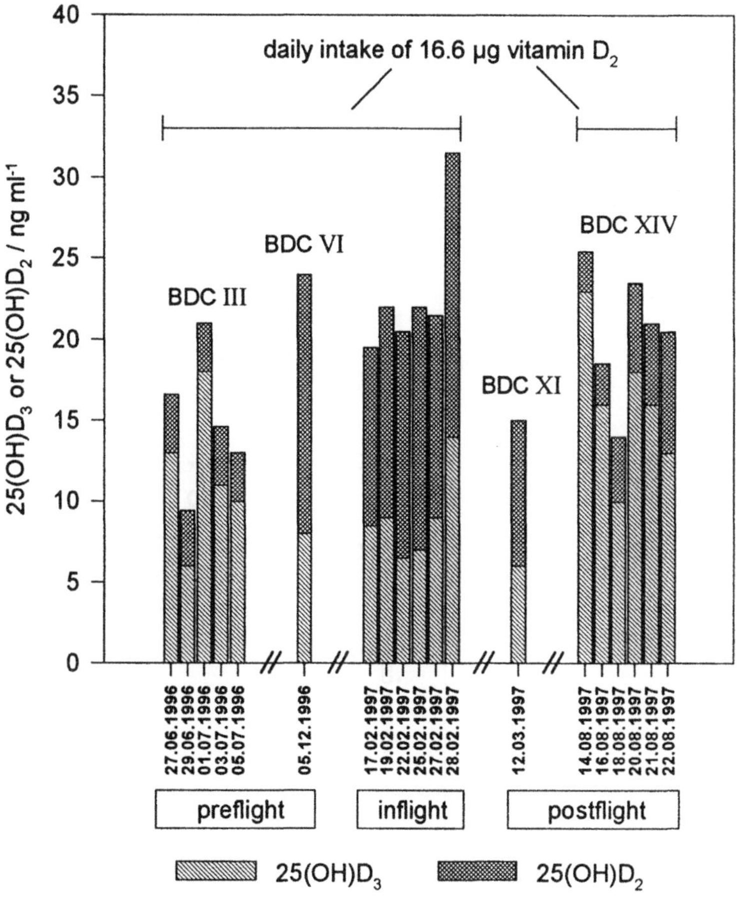

Fig. 5. Serum concentrations of $25(OH)D_3$ and $25(OH)D_2$

DISCUSSION

The biological UV dosimetry onboard the Mir station demonstrated that there was no UV source on board. However, behind the quartz window during direct solar irradiation, a high dose of biologically effective UV radiation was measured even after a few seconds of exposure. This biologically effective UV dose resulted predominantly from the short-wavelength radiation of UVC and UVB since for the exposure intervals choosen in this experiment, the UVA dose was below the limit of detection. This situatiuon is different from the normal scenario on Earth where the Earth's atmosphere cuts-off the short wavelengths of UV mainly through absorption by stratospheric ozone (see Figure 6A). In space the full extra-terrestrial solar spectrum reaches the Mir station. The biological effectiveness of the extra-terrestrial solar radiation and the shielding effects of decreasing ozone concentrations were investigated by using different cut-off filters for simulation of the absorption of different ozone concentrations during the German Spacelab mission D-2 (Horneck et al., 1996). The results of the UVE Experiment on Mir indicate that the quartz window 'illuminator 9' is transparent not only to UVA and visible light but also to UVB and UVC radiation. These short UV wavelengths are the most deleterious to human health as illustrated in the wavelength-dependencies of some of these effects, such as the action spectra for erythema, conjunctivitis, ceratitis, and the induction of skin cancer (in mice) Figure 6B-C. If a cosmonaut is exposed to this radiation behind the quartz window, even for a short period of time, there is a high risk of severe eye and skin damages.

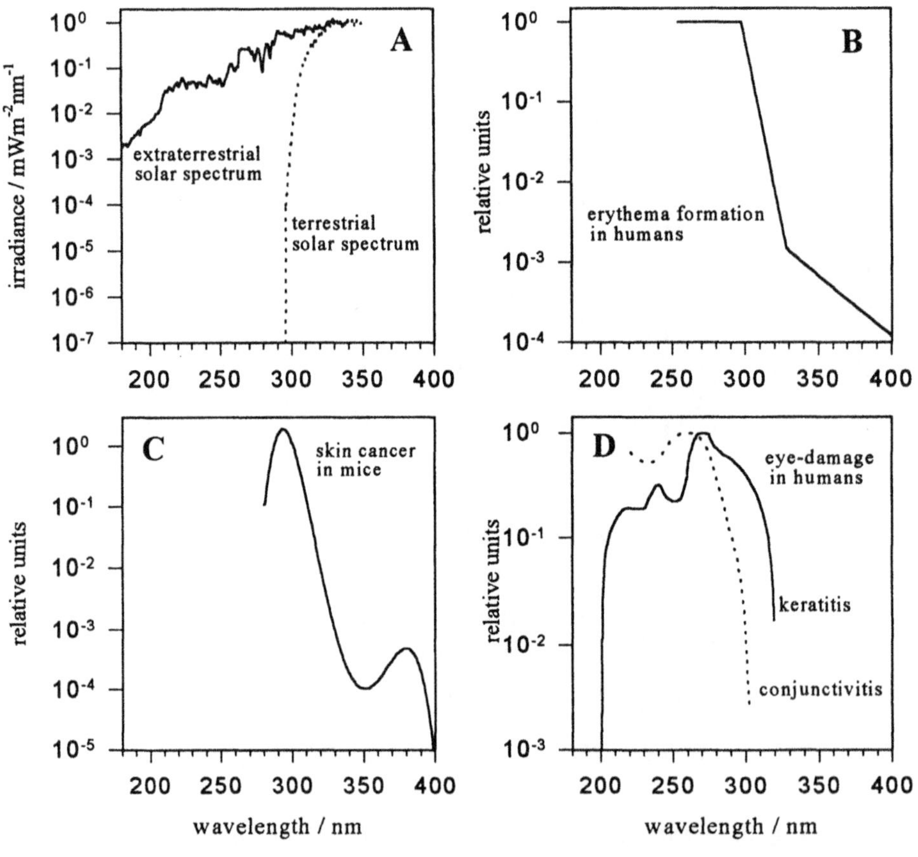

Fig. 6. Solar spectral UV irradiance and action spectra for human health effects
A: extraterrestrial and terrestrial solar UV irradiance (Horneck et al., *1996),*
B: CIE spectrum for erythema formation (McKinley and Diffey, 1987),
C: action spectrum for skin cancer induction in mice (de Gruijl et al., *1993),*
D: action spectra for ceratitis and conjunctivitis (Schreiber and Ott, 1985)

The personal biologically effective UV dose recieved by the cosmonaut during the 15 days of the Mir'97 mission, who was not purposely exposed to solar UV through the quartz window (personal communication) was very low and equivalent to about 1 minimal erythemal dose (MED). Such a dose can be obtained on Earth in about 15 min around noon in the summer months in Europe. In 15 days this dose can not substantially contribute to a sufficient vitamin D supply by cutaneous photosynthesis and will not alter serum 25(OH)D levels (Adams et al., 1982).

Total serum 25(OH)D levels were not reduced during the mission, indicating that vitamin D supply was not involved in alterations of calcium metabolism which were investigated in previous space missions (Rambaut and Johnston, 1979) and during MIR'97 (Zittermann, personal communication). Maintainence of exogenous vitamin D status was the result of high serum 25(OH)D_2 levels due to the vitamin D_2 supplementation. High 25(OH)D_2 serum levels during BDC VI, inflight and during BDC XI are obviously the result of the long term vitamin D supplement in the cosmonaut. The low 25(OH)D_2 levels during the postflight BDC XIV are in line with the fact that vitamin D_2 supplementation was stopped after returning from space for 5 1/2 month and that supplementation was started just a few days prior to the examinations. The reduced inflight serum 25(OH)D_3 levels compared to the postflight BDC XIV values may reflect seasonal variations in cutanous vitamin D synthesis between March and August (McKenna, 1992).

CONCLUSIONS

Taking into account that the results of the UVE Experiment are based on the study of a single subject and that is always a concern whether the observed effects are representative of human population in general the following conclusion can be drawn: (i) Inflight serum levels of vitamin D and its metabolites are not the cause of changes in calcium and bone metabolism observed in microgravity. (ii) Vitamin D supplementation is not necessary in short-term space missions. (iii) For long-term missions a sufficient vitamin D level should be maintained either by controlled nutrient intake of vitamin D and/or by the regular exposure to special UV lamps like those in sun-beds. (iv) The use of natural solar UV radiation through the quartz window for 'sunbathing' is dangerous and should be avoided even for short exposure period.

ACKNOWLEDGMENTS

The space activities were substantially supported by DARA, DLR, the MIR'97 operations team and the scientific project team WPF, which is highly appreciated. The authors especially thank the German cosmonaut R. Ewald.

REFERENCES

Adams J. A. , T. L. Clemens, J. A. Parrish, M. F. Holick, Vitamin-D synthesis and metabolism after ultraviolet irradiation of normal and vitamin-D-deficient subjects, *N. Engl. J. Med.* **306**, 722-725 (1982)

Caillot-Augusseau A., M.-H. Lafage-Proust, C. Solder, J. Pernod, F. Dubois, C. Alexandre, Bone formation and resorption biological markers in cosmonauts during and after a 180-day space flight (Euromir95), *Clin. Chem.* **44**, 578-585 (1998)

de Gruijl F. R., H. J. V. M. Sterenborg, P. D. Forbes, R. E. Davies, C. Cole, G. Kelfkens, H. van Weelden, H. Slaper, J. C. van der Leun, Wavelength Dependence of Skin Cancer Induction by Ultraviolet Irradiation of Albino Hairless Mice, *Cancer Res.* **53**, 53-60 (1993)

Hollis B. W., Comparison of equilibrium and disequilibrium assay conditions for ergocalciferol, cholecalciferol and their major metabolites, *Steroid Biochem.* **21**, 81-86 (1984)

Horneck G., P. Rettberg, E. Rabbow, W. Strauch, G. Seckmeyer, R. Facius, G. Reitz, K. Strauch, J.-U. Schott, Biological dosimetry of solar radiation for different simulated ozone column thicknesses, *J. Photochem. Photobiol. B: Biol.* **32**, 198-196 (1996)

Horneck G., Quantification of the biological effectiveness of environmental UV radiation., *J. Photochem. Photobiol. B: Biol.* **31**, 43-49 (1995)

Maiyar, A. C. and A. W. Norman, Vitamin D, in *Encyclopedia of Human Biology*, Vol. **1**, 859-871, Academic Press, New York (1991)

McKenna M. J., Differences in vitamin D status between countries in young adults and the elderly, *Am. J. Med.* **93**, 69-77 (1992)

McKinley, A. F. and Diffey, B. L., A reference action spectrum for ultraviolet induced erythema in human skin, *CIE J.* **6**, 17-22 (1987)

Oganov V. S., A. I. Grigoriev, L. I. Voronin, A. S. Rakmanonov, A. V. Bakulin, V. Schneider, A. LeBlanc, Bone mineral density in cosmonauts after 4.5 - 6 month long flights aboard orbital station *Mir. Aerospace & Environ. Med.* **26**, 20-24 (1992)

Quintern L. E., G. Horneck, U. Eschweiler, H. Bücker, A biofilm used as ultraviolet-dosimeter, *Photochem. Photobiol.* **55 (3)**, 389-395 (1992)

Patent Nr. DE 40 39 002 A1: Biologische Strahlendetektion; Verfahren und Vorrichtung zur biologischen Detektion von Strahlung durch Mikroorganismen in Form einer auf einem flächigen Substrat befindlichen Mikroorganismen-Beschichtung

Patent Nr. DE 195 35 273: Strahlendosimeter

Rambaut P. C. and R. S. Johnston, Prolonged weightlessness and calcium loss in man, *Acta Astronautica* **6**, 1113-1122 (1979)

Reinhardt, T. A., R. L. Horst, J. W. Orf , B. W. Hollis, A microassay for 1,25-dihydroxyvitamin D not requiring high performance liquid chromatography: application to clinical studies, *J. Clin. Endocrinol. Metab.* **58**, 91-98 (1984)

Rettberg P., G. Horneck, Messung der UV-Strahlenbelastung mit dem DLR-Biofilm, *Derm* **4-1998**, 180-183 (1998)

Schreiber R. and G. Ott, Schutz vor ultravioletter Strahlung, *Schriftenreihe der Bundesanstalt für Arbeitsschutz,* S Nr. 14, Dortmund (1985)

Webb, A. R., Vitamin D Synthsis under Changing UV Spectra, in *Environmental UV Photobiology*, ed. by A. R. Young, Plenum Press, New York (1993)

WHO, Ultraviolet radiation, *Environmental Health Criteria* **160**, WHO, Genf (1994)

Pergamon

Adv. Space Res. Vol. 22, No. 12, pp. 1653–1662, 1998
© 1999 COSPAR. Published by Elsevier Science Ltd. All rights reserved
Printed in Great Britain
0273-1177/98 $19.00 + 0.00

PII: S0273–1177(99)00030–7

BIODOSIMETRY OF HEAVY IONS BY INTERPHASE CHROMOSOME PAINTING

M. Durante, T. Kawata, T. Nakano, S. Yamada and H. Tsujii

National Institute of Radiological Sciences, 9-1-4 Anagawa, Inage-ku, Chiba 263-8555, Japan

ABSTRACT

We report measurements of chromosomal aberrations in peripheral blood lymphocytes from cancer patients undergoing radiotherapy treatment. Patients with cervix or esophageal cancer were treated with 10 MV X-rays produced at a LINAC accelerator, or high-energy carbon ions produced at the HIMAC accelerator at the National Institute for Radiological Sciences (NIRS) in Chiba. Blood samples were obtained before, during, and after the radiation treatment. Chromosomes were prematurely condensed by incubation in calyculin A. Aberrations in chromosomes 2 and 4 were scored after fluorescence *in situ* hybridization with whole-chromosome probes. Pre-treatment samples were exposed *in vitro* to X-rays, individual dose-response curves for the induction of chromosomal aberrations were determined, and used as calibration curves to calculate the effective whole-body dose absorbed during the treatment. This calculated dose, based on the calibration curve relative to the induction of reciprocal exchanges, has a sharp increase after the first few fractions of the treatment, then saturates at high doses. Although carbon ions are 2-3 times more effective than X-rays in tumor sterilization, the effective dose was similar to that of X-ray treatment. However, the frequency of complex-type chromosomal exchanges was much higher for patients treated with carbon ions than X-ray. ©1999 COSPAR. Published by Elsevier Science Ltd.

INTRODUCTION

Human beings on Earth are continuously exposed to radiation. Background radiation is mostly composed of γ-rays, with the exception of the local exposure to α-particles from radon decay in the bronchial epithelium. Photons or β-emitting radionuclides are also used in nuclear medicine. On the other hand, crews of long-term manned space mission are exposed to high-energy heavy ions from galactic cosmic radiation. Although several *in vitro* studies have been performed to elucidate the biological effectiveness of accelerated charged particles, no information is available about the biological consequences of swift heavy nuclei in humans. The space environment contains this unique type of radiation, which has never been experienced by mankind.

Hadrontherapy provides the opportunity now to study the effects of heavy ion exposure in humans. Use of accelerated charged particles in radiotherapy was proposed many years ago, but only recently clinical tests were performed in a large sample of cancer patients at the HIMAC accelerator at NIRS. Carbon ions were selected because of their ballistic (improved dose profile) and radiobiological(high RBE for cell killing and low oxygen-enhancement ratio) properties. Patients treated with accelerated carbon ions represent the first ever experience of mankind to heavy ion exposure on Earth.

Recently, it has been demonstrated that biological dosimetry is a useful technique in assessing the health risk of astronauts due to cosmic-ray exposure (Testard *et al.* 1996, Yang *et al.* 1997, Obe *et al.* 1997). In fact, the quality factor of the mixed charged particle radiation field in space is largely unknown, and the interaction between radiation exposure and microgravity is also unclear. With a biodosimetric test, the "biologically relevant" absorbed dose is measured, and provides a realistic risk assessment. Measurement of reciprocal exchanges in peripheral blood lymphocytes by fluorescence *in situ* hybridization (FISH) with whole-chromosome probes has been proposed as an efficient technique for biodosimetry of space radiation field (Lucas 1997, Durante *et al.* 1997).

Biological dosimetry in patients treated with heavy-ion therapy is therefore interesting for space radiation health risk assessment. Of course, exposure conditions of astronauts and cancer patients are completely different. Low dose, low dose-rate, and whole body exposure is characteristic in space. In contrast, high-dose, daily fractionation at high dose-rate, and partial body exposures are standard in radiotherapy. Nevertheless, useful information about the quality of heavy ion-induce chromosomal damage in lymphocytes can be derived from patient's biodosimetry.

Biodosimetry of cancer patients during radiotherapy treatments is technically challenging. Scoring of dicentrics in peripheral blood lymphocytes at the first mitosis following *in vitro* stimulation demonstrated that in samples obtained during radiation treatment interpretation of the data is rather difficult (Matsubara *et al.* 1974, Liniecki *et al.* 1983, Martin *et al.* 1989, Kleinerman *et al.* 1990, Rigaud *et al.* 1990, Brandan *et al.* 1994, Fong *et al.* 1995). For partial body exposure, interphase death is an important factor affecting measured aberration frequency (Lloyd *et al.* 1973, Ekstrand and Dixon 1982). The problem becomes even more severe for heavy ions, which efficiently induce mitotic delay and interphase death. Premature chromosome condensation (PCC) can be used to overcome these problems: we have shown that scoring chromosomal exchanges in FISH-painted PCC is an accurate biodosimetric test, independent from radiation-induced cell-cycle alterations (Durante *et al.* 1996, 1997). In this paper we report for the first time the results of cytogenetic biodosimetry in patients undergoing X-rays or carbon ion treatment by the PCC+FISH technique.

METHODOLOGY

Premature Chromosome Condensation

The new technique of chemical PCC induction was used in the experiments reported here. This technique is considerable simpler than the conventional fusion method, and provides very high condensation indexes. PCC is induced by incubation in calyculin A following 47 h *in vitro* growth stimulation. The analysis is performed in lymphocytes in different phases of the cell-cycle: G_1, G_2 or M. Results from the different phases are pooled together in the present article. Technical details on chemical PCC induction have been published elsewhere (Durante *et al.* 1998). Briefly, lymphocytes were isolated from whole blood on a Ficoll gradient, and cultured 24 h in RPMI 1640 medium supplemented with 20% fetal calf serum and 1% phytohaemagglutinin. Colcemid at low concentration (40 ng/ml) was then added to the medium, to block the cells in the first post-stimulation cell replication round. After 47 h from the culture initiation, calyculin A (50 nM) was added to the medium and cells were incubated again for 1 h at 37 °C. During this period of time, chromosome condensation was induced in lymphocytes in different phases of the cell-cycle. Finally, lymphocytes were harvested and slides prepared using conventional procedure.

Table 1. Blood Donors for Cytogenetic Monitoring During Radiation Treatment

Patient ID	Age	Sex	Tumor position	Field size# (cm^2)	Radiation	Energy	Dose/ fraction (Gy)$	Fractions/ day
359*	76	M	Upper esophagus	151.5	X-rays	10 MV	1.8	1
183	57	M	Lower esophagus	84	X-rays	10 MV	1.8	1
226	78	M	Middle esophagus	108	C-ions	290 MeV/n	3	1
242	54	F	Lower esophagus	92.3	C-ions	350 MeV/n	2.7	1
191	47	F	Uterus cervix	190	C-ions	350 MeV/n	3	1
161	49	F	Uterus cervix	201	C-ions	350 MeV/n	2.8	1
215	50	F	Uterus cervix	218	X-rays	10 MV	1.5	2

$ - for patients exposed to carbon ions, the dose is expressed in Gy-equivalent (physical dose x RBE in the SOBP); treatment planning is designed to deliver a uniform dose equivalent in the tumor target.
* - tumor was surgically removed and radiation treatment started 2 months after operation.
# - two opposite fields used. Field size can be modified during the treatment.

Chromosome Painting

Slides were hybridized *in situ* with whole-chromosome DNA probes specific for chromosomes 2 (spectrum green) and 4 (spectrum orange), following the protocol suggested by the manufacturer (Vysis, IL, USA) with few modifications (Durante *et al.* 1996). Aberrations were scored with a fluorescent microscope (Olympus) and classified as: reciprocal exchanges (two bi-color chromosomes), including dicentrics and translocations; complex-type exchanges, mostly insertions and non-reciprocal exchanges; and excess acentric fragments. Incomplete exchanges were pooled in the category of reciprocal interchanges, assuming that in most cases the reciprocal fragments were hidden (Wu *et al.* 1998).

In Vitro Irradiation

Lymphocytes isolated at G_0-phase were exposed at room temperature to 200 kVp X-rays at a dose rate of 1 Gy/min. For high-LET experiments, samples were exposed at the HIMAC accelerator to the 290 MeV carbon beam, either with no shielding (average LET= 13 keV/μm) or with 145 mm water-equivalent shielding (average LET including fragments= 83 keV/μm). These correspond to two different position (entrance and distal part) of the spread-out-Bragg-peak (SOBP) beam used in therapy.

Patients

All patients entering the study were informed about its goals and signed an informed consent form. In this report, we compare patients with squamous cell carcinoma of similar size, position, and radiation field size, treated either at the LINAC or HIMAC at NIRS. Details of the patient's treatment profile are reported in Table 1. The physical dose delivered to the tumor was integrated over the target area. The RBE for cell killing of the carbon-beam SOBP varies between 2 and 3.

RESULTS

In Vitro Experiments

Before treatment, lymphocytes from ten patients were exposed *in vitro* to X-rays, to measure individual calibration curves for the induction of chromosomal aberrations. Reciprocal exchanges scored in PCC 2 and 4 are reported in Figure 1. Lymphocyte radiosensitivity *in vitro* was similar for the different patients. Individual calibration curves (linear-quadratic best fit curves displayed in the figure) were used to calculate the effective whole-body dose from the yield of reciprocal exchanges measured *in vivo* during the treatment.

Data in Figure 2 show the *in vitro* biological effectiveness of carbon ions of different LET for the induction of aberrations in PCC 2 and 4. Lymphocytes from the same donor were exposed to the 3 different radiation sources. Compared to X-rays, the carbon beam in the plateau region of the Bragg curve (13 keV/μm) had a similar biological effectiveness, whereas carbon beam at 83 keV/μm was substantially more effective in the induction of all types of aberrations. The RBE, evaluated at an X-ray dose of 3 Gy, was about 2.5 for reciprocal exchanges; about 3 for acentric fragments; and about 8 for complex-type exchanges. The LET of 83 keV/μm corresponds to the distal part of the SOBP, approximately 8.5 mm from the fall-off. The clinical RBE of this portion of the SOBP for cell killing is about 3.

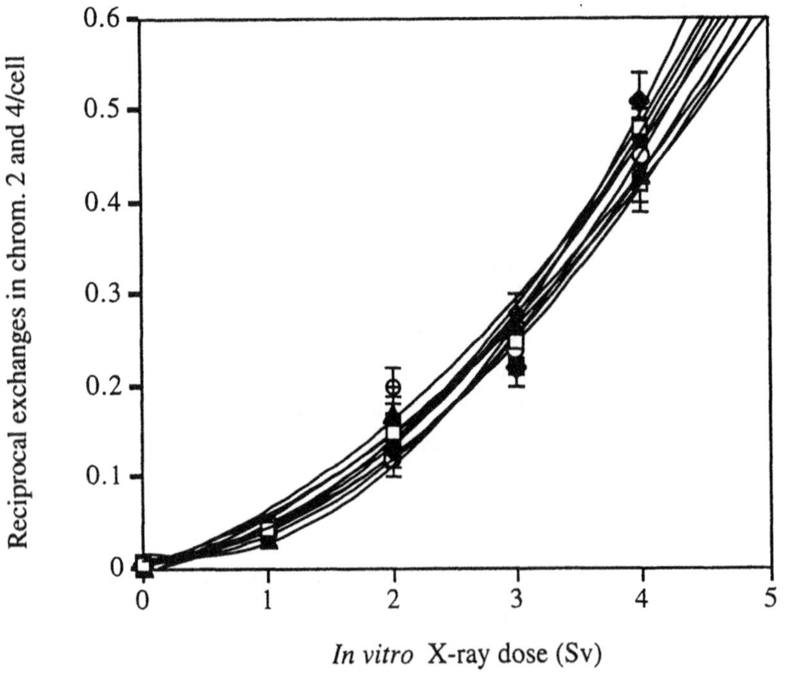

Fig.1. Dose-response curves for the induction of reciprocal exchanges in PCC 2 and 4. Isolated lymphocytes from 10 different patients were exposed *in vitro* to X-rays. Different data symbols refer to different donors. Curves are best fits to the individual patient data point by the function $Y=\alpha D+\beta D^2$ (Y= frequency of reciprocal exchanges in chromosomes 2 and 4, D= X-ray dose).

In Vivo Experiments

Yields of chromosomal aberrations measured during the treatment depends upon the field size and the tumor position. An example is given in Figure 3 for the patient ID no. 359 (see Table 1). The effective whole-body dose was calculated from the yield of reciprocal exchanges by using individual *in vitro* calibration curves (Figure 1). Typical behavior can be observed in the plots: white blood cell and lymphocyte counts decreased after the first week of treatment, and reached a low stable value at high doses. The fraction of aberrant metaphases as well as the yield of chromosomal aberrations per lymphocyte (relative to PCC 2 and 4 only) increased as a function of cumulative tumor dose. In addition, the effective whole-body dose tends to reach a plateau level at high fraction numbers. A negative β-coefficient is generally measured for the dose-response curves *in vivo*, as opposite to the positive β-coefficient observed *in vitro* (Figure 1).

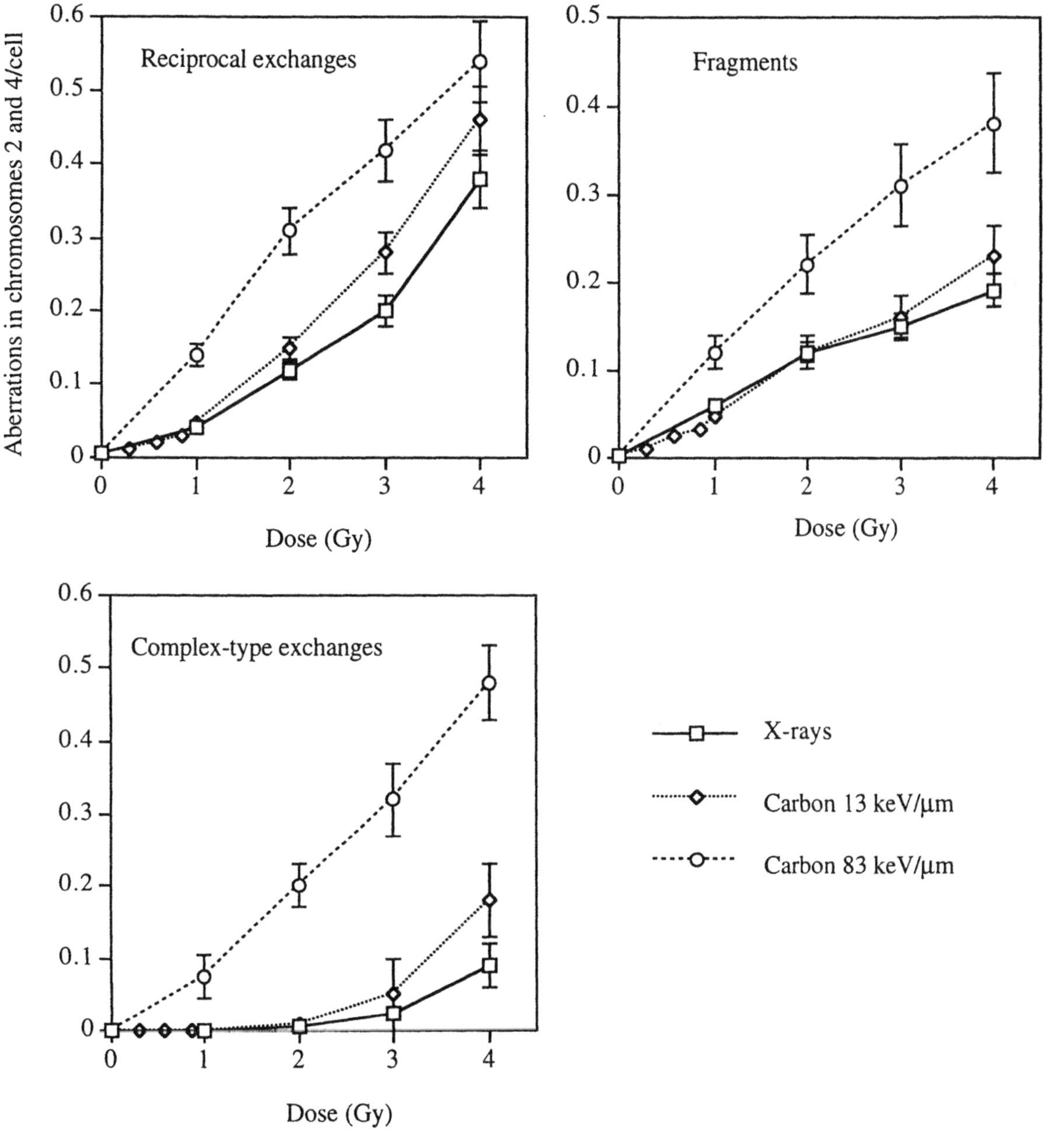

Fig.2. Induction of chromosomal aberrations in peripheral blood lymphocytes from the same donor exposed *in vitro* to either X-rays or ^{12}C-ions. Bars are standard errors of the mean values.

The equivalent whole-body dose in partial body-exposure is sometimes evaluated as the ratio between the intergral dose (kgxGy) and the body weight (Thierens *et al.* 1995). A comparison with our effective doses calculated by using cytogenetic measurements provided a poor correlation, expecially at high doses, when our data indicate a bending of the curve, while the integral dose has a steady increase. A continuous increase in the effective dose is also expected in the mathematical model of Ekstrand and Dixon (1982), who calculated dose distributions to blood cells during external radiotherapy.

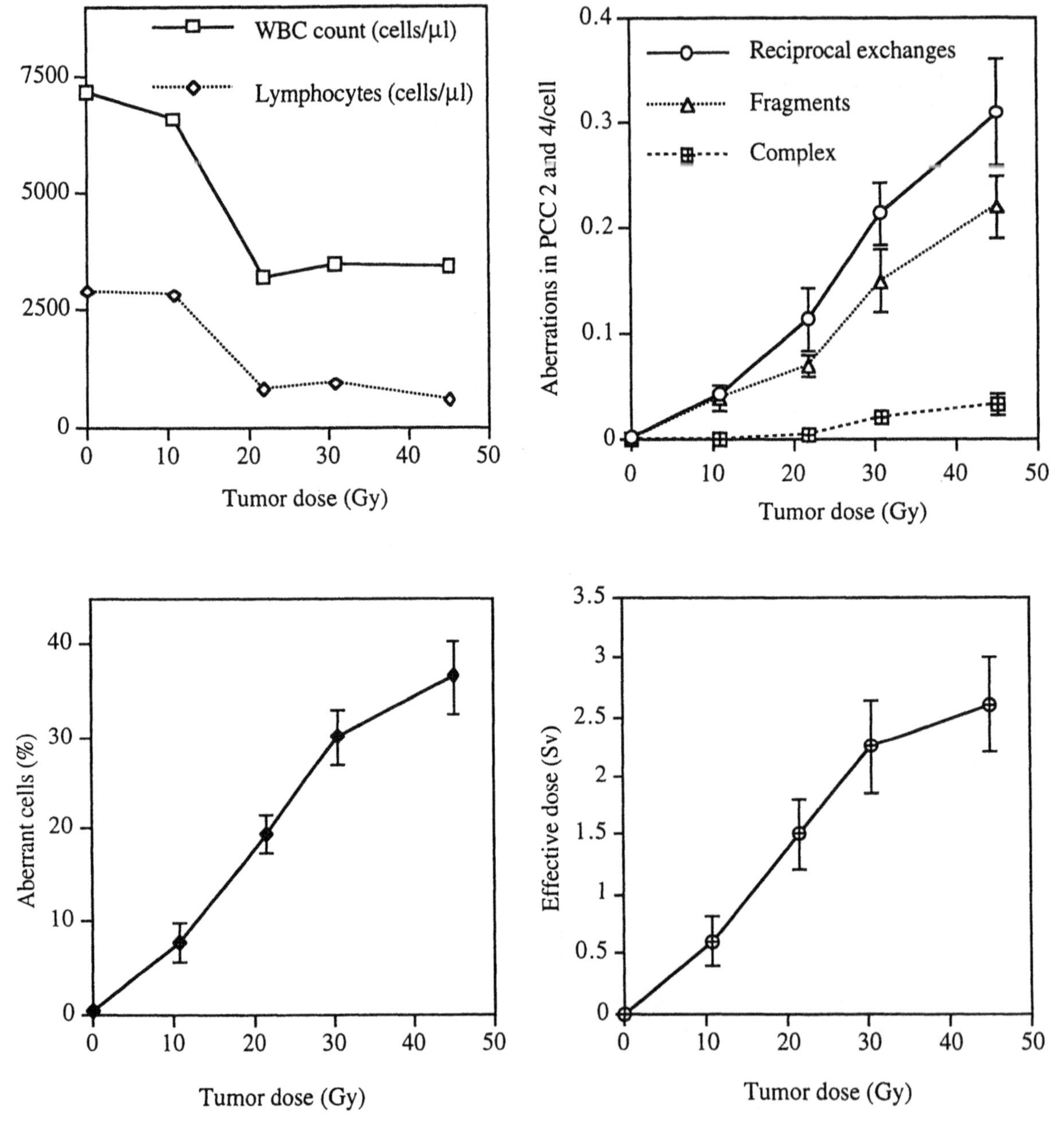

Fig. 3. Chromosome aberrations induced by high-energy X-rays in a patient treated two months after surgical removal of an esophageal carcinoma. Deatils of the patient (ID. no. 359) are given in Table I. Effective dose is calculated by the reciprocal exchanges using individual calibration curves (Figure 1). WBC=white blood cells. Aberrant cells are defined as the fraction of lymphocytes with one or more aberrations involving PCC 2 or 4.

It is likely that the saturation at high doses is caused by repopulation of the lymphocyte pool. Kinetic models of the myelopoietic marrow during fractionated exposure (Jones *et al.* 1991, 1993) might be able to fit these data.

Data relative to the cervical cancer patient ID no. 191 are reported in Figure 4. This patient received a 60 Gy-equivalent (dosexRBE) C-ion radiation dose to treat a cervical cancer.

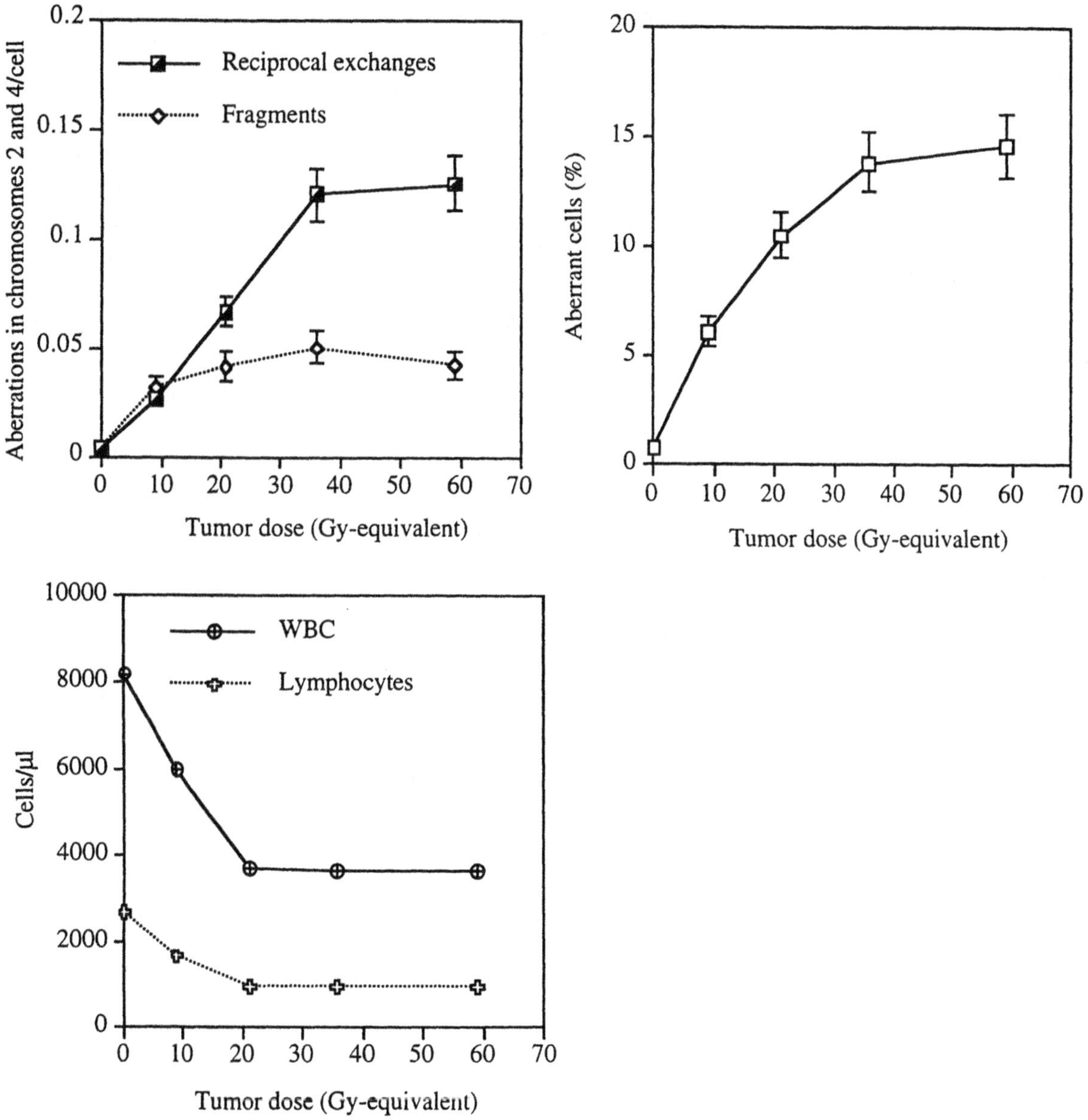

Fig. 4. Dose-response curves for patient no. 191 (see Table I). The equivalent dose to the tumor is reported on the x-axis. For X-rays, 1 Gy-equivalent= 1 Gy. For C-ions, a uniform equivalent dose was delivered to the target volume. The RBE factor varies along the SOBP: for this patient, the value in the middle of the SOBP was 2.4.

The saturation effect at high fraction numbers was comparable to that observed with X-rays. Effective whole-body dose (as calculated from the frequency of reciprocal exchanges using an *in vitro* individual calibration curve) and complex-type exchanges for this patient are displayed in Figures 5 and 6.

Carbon-ions vs. X-rays

The effective whole-body dose can be calculated for patients undergoing photon- and heavy ion-therapy. Results are displayed in Figure 5 for patients with similar tumor sizes and positions. Although the C-ion SOBP is 2-3 times more effective than X-rays in tumor sterilization at the same physical dose delivered to the tumor, no significant differences are observed for the damage of the haemopoietic system. The only exception is patient ID no. 226, treated with C-ions, but for this patient the tumor position is in the middle esophagus (Table 1), which seems to be more sensitive than the lower esophagus (data not shown). Comparison of curves in Figure 3 (X-rays, upper esophagus) and Figure 4 also indicate an increase in lymphocyte damage moving from the abdominal to the cervical esophagus. The relatively low efficiency of C-ions in the induction of cytogenetic damage in the peripheral blood lymphocytes *in vivo* is probably caused by the efficient localization of the physical dose in the tumor region only. A lower fraction of lymph nodes will be exposed with the optimal heavy-ion dose profiles as compared to X-rays. These data suggest that normal tissue can be efficiently spared with hadrontherapy.

However, qualitative differences were observed in the spectrum of chromosomal aberrations induced by C- or X-radiation. *In vitro* data (Figure 2) suggest that C-ions is very effective in the induction of complex-type chromosomal exchanges. *In vivo* data (Figure 6) confirm this observation. These are the first demonstrations that *in vivo* exposure to heavy ions will produce a high fraction of complex exchanges in circulating lymphocytes. The biological significance of these aberrations remain to be elucidated.

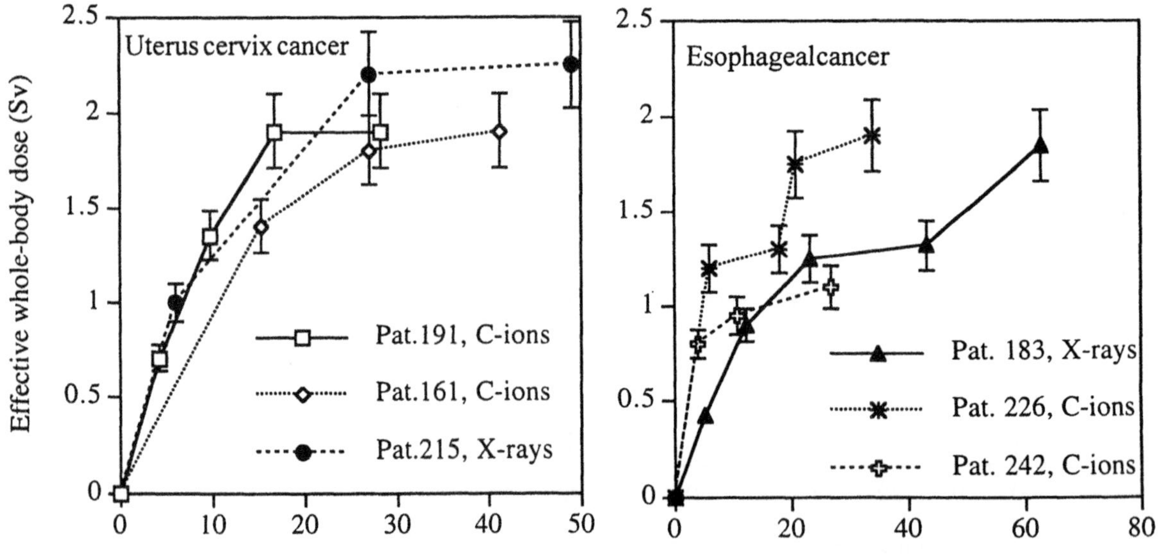

Fig.5. Comparison of the effective whole-body dose (calculated by the measurement of reciprocal exchanges in PCC 2 and 4) absorbed during the treatment of patients with similar tumor size and positions.Left panel, cervical cancer; right panel, esophageal cancer. Bars are standard errors of the mean values.

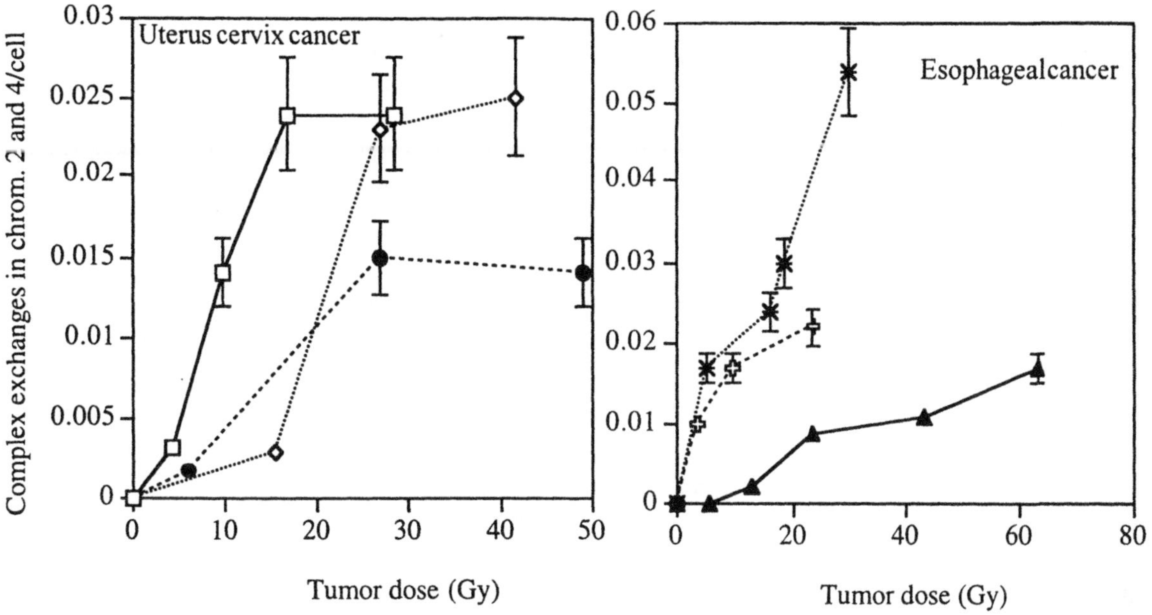

Fig.6. Induction of complex-type chromosomal exchanges in PCC 2 and 4 during the radiation treatment with carbon ions or X-rays. Symbols as in Figure 5. The physical dose delivered to the target area is reported on the x-axis.

CONCLUSIONS

We have shown that interphase chromosome painting (PCC+FISH) is an efficient tool for biodosimetry of heavy ions. Chromosomal aberrations in patients exposed to heavy ions have been reported for the first time. Particularly relevant to space radiation biodosimetry is the observation that heavy ions are very efficient in the induction of complex-type exchanges in blood mononuclear lymphocytes. The ratio complex/simple exchanges might prove to be a useful biomarker of radiation quality.

ACKNOWLEDGMENTS

We are indebted to Dr. Koichi Ando for useful discussions and valuable comments. We also thank Dr. Y.Furusawa and Dr. H.Majima for assistance in HIMAC experiments. M.D. is on leave of absence from the University of Naples "Federico II" (Italy). Financial support from Science and Technology Agency (Japan, fellowship 196121) is gratefully acknowledged.

REFERENCES

Brandan, M.E., M.A. Perez-Pastenes, P. Ostrosky-Wegman, M.E. Gonscbatt, and R. Diaz-Perches, Mean Dose to Lymphocytes During Radiotherapy Treatments, *Health Phys.*, **67**, 326 (1994).

Durante, M., K. George, and T.C. Yang, Biological Dosimetry by Interphase Chromosome Painting, *Radiat. Res.*, **145**, 53 (1996).

Durante, M., K. George, and T.C. Yang, Biodosimetry of Ionizing Radiation by Selective Painting of Prematurely Condensed Chromosomes in Human Lymphocytes, *Radiat. Res.*, **148**, S45 (1997).

Durante, M., Y. Furusawa, and E. Gotoh, A Simple Method for Simultaneous Interphase-Metaphase Analysis in Biological Dosimetry, *Int. J. Radiat. Biol.*, **74**, 457 (1998).

Ekstrand, K.E., and R.L. Dixon, Lymphocyte Chromosome Aberrations in Partial-Body Fractionated Radiation Therapy, *Phys. Med. Biol.*, **27**, 407 (1982).

Fong, L., J-Y. Chen, L-L. Ting, L-T. Lui, P-M. Wang and W-L. Chen, Chromosomal Aberrations Induced in Human Lymphocytes After Partial-Body Irradiation, *Radiat. Res.*, **144**, 97 (1995).

Jones, T.D., M.D. Morris, and R.W. Young, A Mathematical Model of Radiation-Induced Myelopoiesis, *Radiat. Res.*, **128**, 258 (1991).

Jones, T.D., M.D. Morris, and R.W. Young, Mathematical Models of Marrow Cell Kinetics: Differential Effcts of Protracted Irradiation on Stromal and Stem Cells in Mice, *Int. J. Radiat. Oncol. Biol. Phys.*, **26**, 817 (1993).

Kleinerman, R.A., L.G. Littlefield, R.E. Tarone, A.M. Sayer, N.G. Hidreth, L.M. Pottern, S.G. Machad, and J.D. Boyce, Chromosome Aberrations in Relation to Radiation Dose Following Partial-Body Exposure in Three Populations, *Radiat. Res.*, **123**, 93 (1990).

Liniecki, J., A. Bajerska, and K. Wyszynska, Dose-Response Relationships for Chromosome Aberrations in Peripheral Blood Lymphocytes After Whole- and Partial-Body Irradiations, *Mutat. Res.*, **110**, 83 (1983).

Lloyd, D.C., R.J. Purrott, and G.W. Dolphin, Chromosome Aberration Dosimetry Using Human Lymphocytes in Simulated Partial Body Irradiation, *Phys. Med. Biol.*, **18**, 421 (1973).

Lucas, J.N., Dose Reconstruction for Individual Exposed to Ionizing Radiation Using Chromosome Painting, *Radiat. Res.*, **148**, S33 (1997).

Martin, R.H., A. Rademaker, K. Hildebrand, M. Barnes, K. Arthur, T. Ringrose, I.S. Brown, and G. Douglas, A Comparison of Chromosomal Aberrations Induced by *In Vivo* Radiotherapy in Human Sperm and Lymphocytes, *Mutat. Res.*, **226**, 21 (1989).

Matsubara, S., M.S. Sasaki, and T. Adachi, Dose-response Relationship of Lymphocyte Chromosome Aberrations in Locally Irradiated Persons, *Radiat. Res.*, **15**, 189 (1974).

Obe, G., I. Johannes, C. Johannes, K. Hallman, G. Reitz, and R. Facius, Chromosomal Aberrations in Blood Lymphocytes of Astronauts After Long-Term Space Flights, *Int. J. Radiat. Biol.*, **72**, 727 (1997).

Rigaud, O., G. Guedeney, I. Duranton, A. Leroy, M.T. Doloy, and H. Magdelenat, Genotoxic Effects of Radiotherapy and Chemotherapy on the Circulating Lymphocytes of Breast Cancer Patients. I. Chromosome Aberrations Induced *In Vivo*. *Mutat. Res.*, **242**, 17 (1990).

Testard, I., M. Ricoul, F. Hoffschir, A. Flury-Herard, B. Dutrillaux, B.S. Fedorenko, V. Gerasimenko, and L. Sabatier, Radiation-Induced Chromosome Damage in Astronauts' Lymphocytes, *Int. J. Radiat. Biol.*, **70**, 403 (1996).

Thierens, H., A. Vral, M. Van Eijkeren, F. Spelman, and L. De Ridder, Micronucleus Induction in Peripheral Blood Lymphocytes of Patients Under Radiotherapy Treatment For Cervical Cancer or Hodgkin's Disease, *Int. J. Radiat. Biol.*, **67**, 529 (1995).

Wu, H., K. George, and T.C. Yang, Estimate of True Incomplete Exchanges Using Fluorescence *In Situ* Hybridization with Telomere Probes, *Int. J. Radiat. Biol.*, **73**, 521 (1998).

Yang, T.C., K. George, A.S. Johnson, M. Durante, and B.S. Fedorenko, Biodosimetry Results from Space Flight MIR-18, *Radiat. Res.*, **148**, S17 (1997).

Pergamon

Adv. Space Res. Vol. 22, No. 12, pp. 1663–1671, 1998

Printed in Great Britain
0273-1177/98 $19.00 + 0.00

PII: S0273-1177(99)00031-9

RESIDUAL CHROMATIN BREAKS AS BIODOSIMETRY FOR CELL KILLING BY CARBON IONS

M.Suzuki[1], Y.Kase[1], T.Nakano[2], T.Kanai[3] and K.Ando[1]

1 *Space and Particle Radiation Science Research Group, National Institute of Radiological Sciences, 4-9-1 Anagawa, Chiba-shi 263, Japan*

2 *Division of Radiation Medicine, National Institute of Radiological Sciences, 4-9-1 Anagawa, Chiba-shi 263, Japan*

3 *Division of Accelerator Physics and Engineering, National Institute of Radiological Sciences, 4-9-1 Anagawa, Chiba-shi 263, Japan*

ABSTRACT

We have studied the relationship between cell killing and the induction of residual chromatin breaks on various human cell lines and primary cultured cells obtained by biopsy from patients irradiated with either X-rays or heavy-ion beams to identify potential bio-marker of radiosensitivity for radiation-induced cell killing. The carbon-ion beams were accelerated with the Heavy Ion Medical Accelerator in Chiba (HIMAC). Six primary cultures obtained by biopsy from 6 patients with carcinoma of the cervix were irradiated with two different mono-LET beams (LET= 13 keV/μm, 76 keV/μm) and 200kV X rays. Residual chromatin breaks were measured by counting the number of non-rejoining chromatin fragments detected by the premature chromosome condensation (PCC) technique after a 24 hour post-irradiation incubation period. The induction rate of residual chromatin breaks per cell per Gy was the highest for 76 keV/μm beams on all of the cells. Our results indicated that cell which was more sensitive to the cell killing was similarly more susceptible to induction of residual chromatin breaks. Furthermore there is a good correlation between these two end points in various cell lines and primary cultured cells. This suggests that the detection of residual chromatin breaks by the PCC technique may be useful as a predictive assay of tumor response to cancer radiotherapy. ©1999 COSPAR. Published by Elsevier Science Ltd.

INTRODUCTION

The premature chromosome condensation (PCC) technique is a very powerful method for detecting chromatin damage in G_1- or G_2-phase cells. Induction of chromatin damage detected by the PCC technique in cells irradiated with either low- or high-LET radiations showed a linear dose response in the same dose range for cell-survival experiment (Pantelias and Maillie 1985, Bedford and Goodhead 1989, Goodwin *et al.* 1989, Cornforth and Goodwin 1991, Suzuki *et al.* 1992, Loucas and Geard

1994,Goodwin *et al.* 1994, Suzuki *et al.* 1996). Most of the induced chromatin breaks were rejoined within a few hours after irradiation (Hittelman and Rao 1974, Cornforth and Bedford 1983, Iliakis and Pantelias 1990, Suzuki *et al.* 1992, Goodwin *et al.* 1994).

Using the PCC technique, some investigators have shown that radiation-induced residual chromatin breaks in mammalian cells have a good correlation with its cell killing effect (Pantelias and Maillie 1984, Goodwin *et al.*1989, Brown *et al.*1992, Pandita and Hittelman 1992, Sasai *et al.* 1994, Suzuki *et al.* 1996, Suzuki *et al.* 1997). Fluorescence in situ hybridization in combination with the PCC technique has been assessed as potential predictive assay for radiotherapy of tumor cells and as biological dosimeter in irradiated human lymphocytes (Brown *et al.* 1992, Coco-Martin *et al.*1994, Sasai *et al.* 1994, Durante *et al.* 1996).

In the present study, we examined cytotoxicity and induction of residual chromatin breaks in primary cervical carcinoma cells irradiated with different types of radiation such as X rays and heavy-ion beams. Tumor cultures were cells obtained by biopsy from 6 patients treated with the HIMAC. Results were compared with the data previously obtained using various human cell lines (Suzuki *et al.* 1998). We show here that PCC technique is a reliable biological marker of the intrinsic radiosensitivity of tumor cells and the results are comparable to clonogenicity assay.

MATERIALS AND METHODS

Cells

Six primary cultured cells obtained by biopsy from 6 cervical carcinoma patients who were treated with the Heavy Ion Medical Accelerator in Chiba (HIMAC) were used as in the study. Punched biopsies of about 4 to 7 mm diameter were placed in phosphate-buffered saline (PBS) immediately after removal from patients. Each biopsy was then decontaminated by brief treatment in 70% ethanol.The biopsy was minced into small fragments in a plastic dish (Falcon 3003) and treated with 0.2% trypsin in a polypropylene tube (Falcon 2059) for 5 min at 37°C. Eagle's minimum essential medium (MEM) supplemented with 10% fetal bovine serum was added to inactivate the trypsin and the samples were washed twice in PBS. The final cell suspension was cultured in MEM with 10% fetal bovine serum in a 5% CO_2 incubator at 37°C. The growing cell population was called passage 0 and representative cultures were frozen in liquid nitrogen and the obtaining cells sub-cultured to expanded to be used for our study. We used confluent monolayers of cells at passage 3 to 5 and the chromosome number of each biopsied cell was stable. The chromosome number (average $\pm$ standard deviation) of each of the 6 primary cultured cells was 44.0 $\pm$ 1.7 for "a" cell, 47.2 $\pm$ 2.2 for "b" cell, 47.3 $\pm$ 2.4 for "c" cell, 46.8 $\pm$ 2.3 for "d" cell, 45.9 $\pm$ 1.1 for "e" cell and 46.1 $\pm$ 1.2 for "f" cell, respectively. All of these cell lines were fibroblastic in morphology when examined under the microscope. These cells were not likely to be squamous cell carcinoma, since they did not secrete the tumor markers of squamous cell carcinoma (SCC) into the medium (data not shown). Mitotic XP2OS cells were used as inducers of PCC. This cells also were cultured in MEM supplemented with 10% fetal bovine serum in a 5% CO_2 incubator at 37°C.

Irradiation

The 6 primary target cells were irradiated with carbon-ion beams accelerated by HIMAC. Details concerning the HIMAC beam-delivery system, physical characteristics, irradiation procedures and dosimetry have been described elsewhere (Kanai *et al.* 1994). Briefly, the initial energy of the carbon-ion beams was 290 MeV/n. We used two kinds of beams having different LET values using Lucite absorbers with various thicknesses to change the energy of the beams. At the sample position, we estimated that the LET_∞ values to be 13.3 keV/μm for low-LET beams and 76.4±0.9 keV/μm for high-LET beams. The dose rate of both LET beams was about 1.2 Gy/min. For comparison, we used 200 kV X rays at a dose rate of 0.85 Gy/min. All of the irradiations were carried out at room temperature.

Cell-survival assay

After irradiation, cells were immediately plated onto 60mm (Falcon 3002) or 100mm (Falcon 3003) plastic dishes at a density estimated to form 60 to 70 viable colonies per dish for cell-survival assay. Colonies were fixed and stained with 20% methanol and 0.2% crystal violet (Wako Pure Chemical Industries Ltd.) after a 14-day incubation period. Colonies consisting of more than 50 cells were scored as a surviving colony. The plating efficiency of the 6 cells ranged from 12 to 56% in this study.

PCC assay

The irradiated target cells were incubated for 24 hr in a 5% CO_2 incubator to allow rejoining of induced PCC breaks. Residual chromatin breaks were measured as remaining fragments of prematurely condensed chromosomes. The procedure for the induction of premature chromosome condensation has been reported in detail elsewhere (Pantelias and Maillie 1983, Suzuki *et al.* 1996). Briefly, $1x10^6$ mitotic XP2OS cells were used as PCC inducer, produced by a 6hr-incubation period in the presence of 0.1μg/ml demecolcine (Wako Pure Chemical Industries Ltd.), were mixed with an equal number of irradiated target cells in a polypropylene tube (Falcon 2059). After centrifugation at 200g for 5min, the pellet cell was exposed to 0.15ml of 50%(w/v) polyethyleneglycol (PEG; M.W.=1540, Boehringer Mannheim GmbH, Germany) in 75mM Hepes buffer for 1min. Four milliliters of PBS was then gently added to the tube and the cell suspension was centrifuged at 200g for 5min. The pellet was resuspended in 5ml MEM medium containing 0.1μg/ml demecolcine and incubated in a CO_2 incubator for 1hr. These cells were subsequently treated with a 75mM KCl solution for 20min at room temperature and fixed with Carnoy's solution (methanol : acetic acid = 3:1). The cell suspension was dropped onto slides, air dried, and stained with a 5% Giemsa solution. PCC preparations of 20 cells were scored under a light microscope. The yield of PCC breaks per cell was estimated to be the number of PCC fragments in excess of the number of PCC fragments found in the non-irradiated cells.

RESULTS AND DISCUSSION

Survival curves for 6 primary cultured cells

Figure 1 shows the dose-response survival curves of the 6 primary cervical carcinoma cultures irradiated with either X rays or low- and high-LET carbon-ion beams. The results indicated that primary cultures which were sensitive to X rays were similarly sensitive to both low- and high-LET carbon-ion beams. The difference in radiosensitivity for high-LET beams among the 6 primary cultures, however, was smaller than that for X rays. The D_{10} and RBE values relative to X rays were summarized in table 1. The RBE values ranged from 1.1 to 1.3 for the low-LET beam and 1.8 to 2.7 for the high-LET beam in this study.

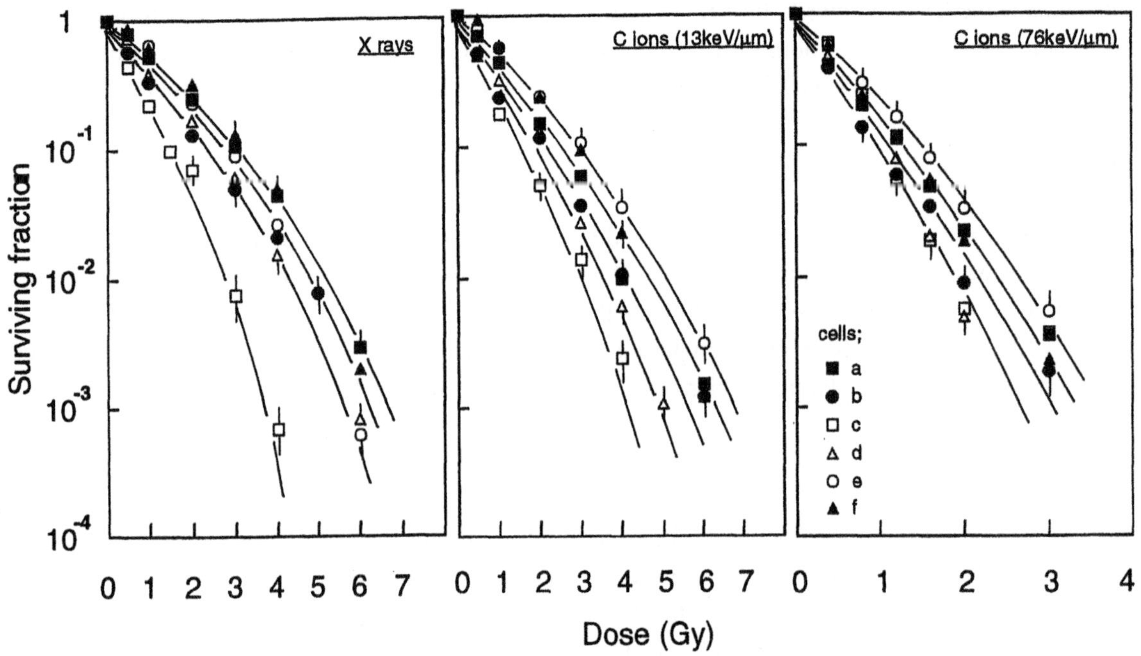

Fig.1 Dose-response survival curves of 6 primary cervical carcinoma cell cultures irradiated with X rays, low-LET (13.3 keV/µm) and high-LET (76.4±0.9 keV/µm) carbon-ion beams. The data points show the mean values and standard error of at least three independent experiments in each cell lines and each radiation types. The curves are fitted by the eye.

Table 1. D10 and RBE values for cell killing effect in 6 kinds of primary cultured cells obtained by biopsy irradiated with X rays and carbon-ion beams.

Cells	X rays	C ions (13.3keV/µm)		C ions(76keV/µm)	
	D_{10} (Gy)	D_{10} (Gy)	RBE	D_{10} (Gy)	RBE
a	3.15	2.42	1.30	1.15	2.74
b	2.56	1.94	1.32	1.03	2.49
c	1.70	1.50	1.13	0.96	1.77
d	2.57	2.13	1.21	1.13	2.27
e	3.24	2.89	1.12	1.42	2.28
f	2.86	2.67	1.07	1.32	2.17

Table 2. The number of residual chromatin breaks per Gy on 6 primary cultures by X rays and C ions

Cell	No. of non-rejoining PCC breaks per Gy		
	X rays	C ions (13keV/μm)	C ions (76keV/μm)
a	1.3	3.3	7.4
b	1.6	3.5	7.5
c	1.9	4.0	8.0
d	1.6	3.3	6.7
e	1.4	2.9	6.3
f	1.3	3.0	5.4

Figure 3 shows the number of residual chromatin breaks per cell per Gy for X rays, the low-LET and the high-LET carbon-ion beams as a function of their D10 values.The data for using the human cell lines was previously reported by Suzuki *et al.* (1998). This result indicated that there was a good correlation between radiosensitivity for cell killing effect and the induction of residual chromatin breaks in primary cultured cells as well as established cell lines. Pantelias and Maillie (1984) reported the use of PCC technique in human blood mononuclear cells exposed to X rays as a biological dosimeter and several investigators also showed that the radiosensitivity of human cells irradiated with photons and heavy ions had a good correlation with the induction of residual chromatin breaks (Goodwin *et al.* 1989, Cornforth and Goodwin 1991, Pandita and Hittelman 1992, Suzuki *et al.* 1996, Suzuki *et al.* 1997). Furthermore, PCC technique combined with fluorescence *in situ* hybridization (FISH) has been used as predictive assay of intrinsic radiosensitivity of tumor cells in cancer radiotherapy (Brown *et al.* 1992, Coco-Martin *et al.* 1994, Sasai *et al.* 1994, Durante *et al.* 1996). The advantage of using the PCC technique in predicting radiosensitivity of tumor cells are 1) useful for non-cycling / non-clonogenic cells, 2) faster in obtaining the result than using colony-forming assay, and 3) being a direct induction of repair of radiation damage. It is, therefor, very useful for biopsy samples to use the PCC technique. The data obtain by the PCC technique may accurately reflect the radiation damage for the chromatin level without the competition as a result of cell cycle delay and/or interphase cell death. This means that the PCC technique is very useful for detecting chromatin damages by qualitatively different types of radiation. Our data show the possibility of using the PCC technique as one of the biological markers of intrinsic radiosensitivity of tumor cells and as a predictive assay for clinical radiotherapy. It may be necessary to use several different kinds of the biological assays, such as the PCC, the micronucleus and the MTT assay to obtain a precise prediction of the intrinsic radiosensitivity of tumor cells.

Induction of remaining chromatin breaks

Figure 2 shows dose-response curves for the induction of residual chromatin breaks after 24hr of post-irradiation incubation in the 6 primary cell cultures. The number of residual chromatin breaks per Gy was summarized in table 2. The frequency for the induction of residual chromatin breaks by high-LET carbon-ion beams was the highest of the 3 different kinds of radiation used in this study. The number of residual chromatin breaks per Gy ranged from 1.3 to 1.9 for X rays, 2.9 to 4.0 for the low-LET carbon-ion beams and 5.4 to 8.0 for the high-LET carbon-ion beams. The cell line "c" which was most radiosensitive cell in this study based on clonogenic survival showed the highest induction of residual chromatin breaks by all 3 radiations examined.

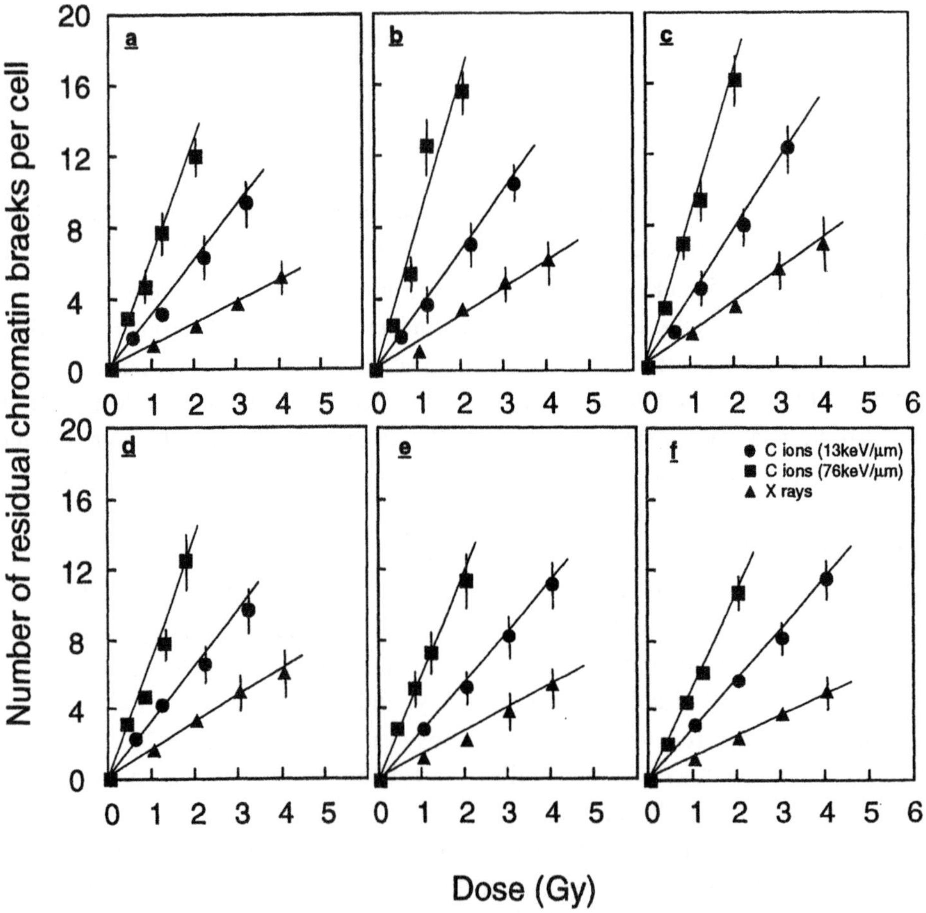

Fig.2 Dose-response curves for the induction of residual chromatin breaks after 24 hr of post-irradiation incubation by X rays (▲), low-LET (●) and high-LET (■) carbon-ion beams in 6 primary cervical carcinoma cell lines (a through f). The data points show the mean values and standard deviation of two independent experiments.

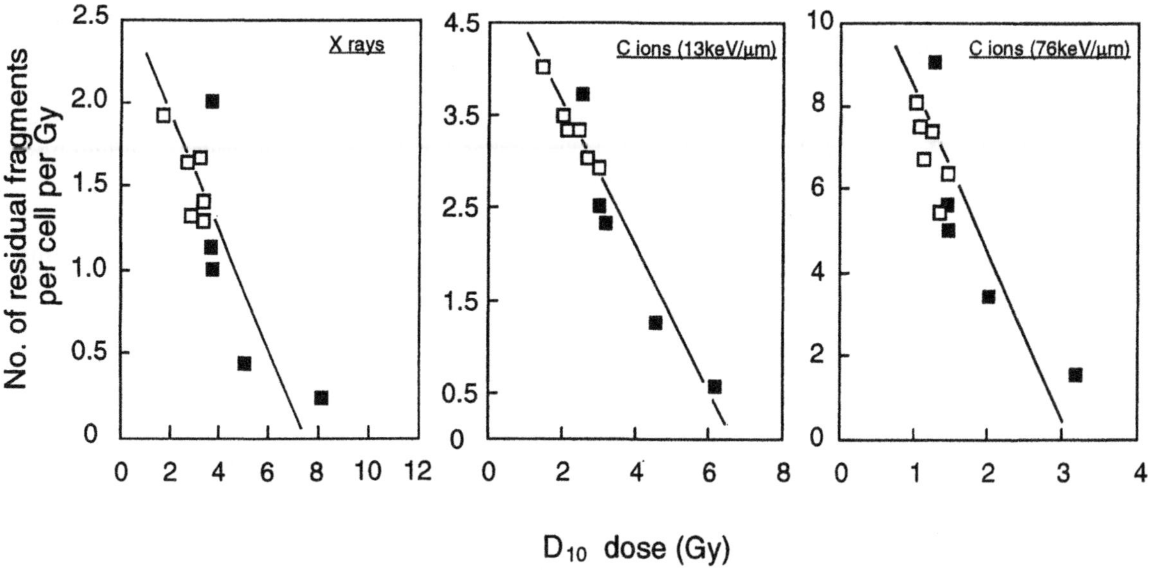

Fig.3 Number of residual PCC breaks per cell per Gy for X rays, the low-LET and the high-LET carbon-ion beams as a function of their D10 values for 6 primary cervical carcinoma cell cultures obtained (open square symbols) and 5 different human cell lines which are 2 kinds of normal human fibroblast cells and 3 kinds of human tumor cell lines (closed square symbols). The data of the human cell lines are taken from Suzuki *et al.* (1998).

ACKNOWLEDGMENTS

We thank the staff of HIMAC for help with the irradiation using carbon-ion beams. This work was supported by the Research Project with Heavy Ions at NIRS-HIMAC.

REFERENCES

Bedford, J.S. and Goodhead, D.T., Breakage of human interphase chromosomes by alpha particles and X-rays, *Int. J. Radiat. Biol.*, **55**, 211 (1989).

Brown, J.M., J.Evans, and M.S.Kovacs, The prediction of human tumor radiosensitivity *in situ*: An approach using chromosome aberrations detected by fluorescence *in situ* hybridization, *Int.J.Radiation Oncology Biol.Phys.*, **24**, 279 (1992).

Coco-Martin, J.M., M.F.M.A.Smeets, M.Poggensee, E.Mooren, I.Hofland, *et al.*, Use of fluorescence *in situ* hybridization to measure chromosome aberrations as a predictor of radiosensitivity in human tumour cells, *Int. J. Radiat. Biol.*, **66**, 297 (1994).

Cornforth, M.N., and J.S. Bedford, X-ray-induced breakage and rejoining of human interphase chromosomes, *Science*, **222**, 1141 (1983).

Cornforth, M.N. and Goodwin, E.H., The dose-dependent fragmentation of chromatin in human fibroblasts by 3.5-MeV a particles from ^{238}Pu: Experimental and theoretical considerations pertaining to single-track effects, *Radiat. Res.*, **127**, 64 (1991).

Durante, M., K.George, and T.C.Yang, Biological dosimetry by interphase chromosome painting, *Radiat. Res.*, **145**, 53 (1996).

Goodwin,E., E.Blakely, G.Ivery, and C.Tobias, Repair and misrepair of heavy-ion-induced chromosomal damage, *Adv. Space Res.*, **9**, (10)83 (1989).

Goodwin, E.H., E.A.Blakely, and C.A.Tobias, Chromosomal damage and repair in G1-phase Chinese hamster ovary cells exposed to charged-particle beams, *Radiat. Res.*, **138**, 343 (1994).

Hittelman,W.N., and P.N.Rao, Premature chromosome condensation. I.Visualization of X-ray induced chromosome damage in interphase cells, *Mutat.Res.*, **23**, 251 (1974).

Iliakis, G.E., and G.E. Pantelias, Production and repair of chromosome damage in an X-ray sensitive CHO mutant visualized and analysed in interphase using the technique of premature chromosome condensation, *Int. J. Radiat. Biol.*, **57**, 1213 (1990).

Kanai, T., Tomura, H., Matsufuji, N., Minohara, S., Fukumura, A.,*et al.*, HIMAC beam delivery system -Physical characteristics-, *Proceedings of NIRS International Seminar on the Application Therapy of Cancer in connection with XXI PTCOG Meeting*, 26 (1994).

Loucas, B.D. and Geard, C.R., Initial damage in human interphase chromosomes from alpha particles with linear energy transfers relevant to Radon exposure, *Radiat. Res.*, **139**, 9 (1994).

Pantelias, G.E. and Maillie, H.D., A simple method for premature chromosome condensation induction in primary human and rodent cells using polyethyleneglycol, *Somatic Cell Genetics*, **9**, 533 (1983).

Pantelias, G.E., and H.D. Maillie, The use of peripheral blood mononuclear cell prematurely condensed chromosomes for biological dosimetry, *Radiat. Res.*, **99**, 140 (1984).

Pantelias, G.E. and Maillie, H.D., Direct analysis of radiation-induced chromosome fragments and rings in unstimulated human peripheral blood lymphocytes by means of the premature chromosome condensation technique, *Mutat.Res.*, **149**, 67 (1985).

Pandita, T.K. and Hittelman, W.N., The contribution of DNA and chromosome repair deficiencies to the radiosensitivity of Ataxia-Telangiectasia, *Radiat. Res.*, **131**, 214 (1992).

Sasai, K., Evans, J., Kovacs, M.S. and Brown, J.M., Prediction of human cell radiosensitivity: comparison of clonogenic assay with chromosome aberrations scored using premature chromosome condensation with fluorescence *in situ* hybridization, *Int.J.Radiation Oncology Biol.Phys.*, **30**,1127 (1994).

Suzuki,M., M.Watanabe, K.Suzuki, K.Nakano, and K.Matsui, Heavy ion-induced chromosome breakage studied by premature chromosome condensation (PCC) in Syrian hamster embryo cells, *Int. J. Radiat. Biol.*, **62**, 581 (1992).

Suzuki,M., M.Watanabe, T.Kanai, Y.Kase, F.Yatagai, T.Kato, *et al.*, LET dependence of cell death, mutation induction and chromatin damage in human cells irradiated with accelerated carbon ions, *Adv. Space Res.*, **18**, (1/2)127 (1996).

Suzuki,M., Y.Kase, T.Kanai, F.Yatagai, and M.Watanabe, LET dependence of cell death and chromatin-break induction in normal human cells irradiated by neon-ion beams, *Int. J. Radiat. Biol.*, **72**, 497 (1997).

Suzuki,M., Y.Kase, T.Kanai, and K.Ando, Correlation between cell death and induction of non-rejoining PCC breaks by carbon-ion beams, *Adv. Space Res.*, **22**, 561 (1998).

Pergamon

Adv. Space Res. Vol. 22, No. 12, pp. 1673–1682, 1998

Printed in Great Britain
0273-1177/98 $19.00 + 0.00

PII: S0273-1177(99)00032-0

INDUCTION OF ASYMMETRICAL TYPE OF CHROMOSOMAL ABERRATIONS IN CULTURED HUMAN LYMPHOCYTES BY ION BEAMS OF DIFFERENT ENERGIES AT VARYING LET FROM HIMAC AND RRC.

H. OHARA[1], N.OKAZAKI[1], M. MONOBE[1], S. WATANABE[1], M. KANAYAMA[1], and M. MINAMIHISAMATSU[2]

1 *Department of Biology, Faculty of Science, Okayama University ; 3-1-1 Tsushima-naka, Okayama 700-8530, Japan*
2 *Division of .Radiobiology and Biodosimetry, National Institute of Radiological Sciences; 4-9-1 Anagawa, Chiba 263-8555, Japan*

ABSTRACT

Frequencies of asymmetrical type of chromosome aberration were scored in cultured human blood lymphocytes irradiated with carbon and neon beams. Blood cells were irradiated with various doses to establish dose response curves for chromosome aberration frequency vs. dose, and chromosome preparation was made by conventional method. Dose response curves for the per cell frequencies of the dicentrics and centric rings as well as the excess amount of acentric fragments were described for 7 different qualities (LET= 22.4, 40.0, 41.5, 69.9, 70.0, 100.0 and 150 KeV/μm) of carbon and neon beams with three different energies, 135, 290 and 400 MeV/u. From the analysis of those dose response curves, the maximum effect was found in the region of LET value at near 70 KeV/μm together with linear expression in the response from all endpoints examined. The 135 MeV/u of carbons (69.9 KeV/μm) and neons(70.0 KeV/μm) showed linear response. The 290 MeV/u of carbons (100 KeV/m) and neons (150 KeV/μm) showed medium effects with different shape of response, linear with a plateau and upward concavity. The 2 carbon beams (41.5 and 40 KeV/μm) from 2 different accelerators showed much discrepancy in the response. RBE-LET relationship was also described by comparing the coefficient α of the 7 different dose responses.The peak (near 70 KeV/m) was localized close to that (80 KeV/m) for the survivals of dsb repair deficient cells (Eguchi-Kasai et al. 1998), but in different position from that previously reported in many other studies (100-200 KeV/mm). Identification of the RBEmax in the present study has yet to be definitive.

INTRODUCTION

The advance in the field of heavy ion radiobiology has been endorsed by the development in large-scaled accelerator constructions, which were achieved in LBL (USA), GSI (Germany), Riken (Japan) and NIRS (Japan). Among them, the 2 Japanese heavy ion accelerators, Riken ring cyclotron (RRC) for heavy ions and HIMAC (Heavy ion medical accelerator in Chiba) are the latest facilities available for studies in physics, biology and medicine since 1992. The HIMAC facility is designed for medical purposes and mainly directed at the treatments of malignant cancer as well as for studies in radiation physics, technology and biology . The present study was carried out for the analysis of chromosome aberrations induced by those particle ion beams produced by the operations of these 2 Japanese heavy ion accelerators with 7 different qualities in LET (Linear Energy Transfer).

The study of chromosome damage provides useful information about the mechanism of radiation action at the cellular level. Recent cytological technique allows investigators to detect the number of lesions and their distributions on a per cell basis. In studies with irradiated human lymphocytes, the measurement of chromosomal aberrations provides the most sensitive method for biological dosimetry. The establishment, therefore, of the dose response for induction of chromosome aberrations is also important not only for calibrating average whole body dose of accidental exposure but also for the prediction of radiation effects due to undesired radiation exposures. Chromosome study on higher LET radiations will make it possible to predict radiation exposure to be encountered in space environment for the strategy of radiation protection. In measuring of chromosome aberrations, the dicentrics has been conveniently used mainly because it involves an exchange between two chromosomes, and is considered as the one most reliably scored (Edwards 1997). However, measurements of translocation are only possible by using modern technique of chromosome painting. For this reason and another, only those of asymmetrical type of chromosome aberrations were scored in the present study.

During last 3 decades, a hump-shaped response curve for the fluctuation of RBE vs. LET has been demonstrated for many end points including chromosome aberrations, reproductive death, mutation inductions, DNA breaks and lethal damages in mammalian cells (Skarsgard et al. 1967, Barendsen 1993 and 1997, Blakely et al. 1984, Kraft 1987, Furusawa et al. 1992). In view of these accumulated cell biological data, the maximum effects could be identified as the optimal ionization density for chromosome aberrations. In the present study, the total of 7 different LET of particle radiations, 5 different carbon and 2 different neon beams were investigated . The results indicate that higher LET particle beams near the maxima for aberration induction enhance the excess induction of acentric fragments, suggesting the prevention of chromosome rejoining and that high dose irradiation with higher LET induces a more complex structural change in chromosome like polycentric type of aberration.

MATERIALS AND METHODS

Blood Culture

Human heparinized blood was obtained from a healthy male donor and kept at room temperature. Blood was then transfered into a series of irradiation flasks made of plexiglass which were specially designed for the purpose of particle beam irradiation. Flasks were filled to capacity and were set in the irradiation apparatus for exposure at room temperature. Under these conditions blood lymphocytes were considered to be in Go phase. After irradiation, blood samples were distributed into culture flasks containing RPMI-1640 medium with 20 % fetal calf serum, phytohaemagglutinin (20 μl/ml) and colcemid (0.01μg/ml). Metaphases were harvested mostly from 53-h cultures but partly from 60-h cultures because of division delay induced by high doses and high LET effects of irradiation as well. Metaphase cells harvested were considered to be in the first post-irradiation mitosis because colcemid was present in cultures from the start to prevent the 2nd mitosis.

Chromosome Preparation

Preparation was made according to standard techniques (IAEA 1986). Cells in the cultures at the time of harvest were collected by centrifugation, treated with hypotonic solution (0.075M KCl) and fixed with methanol-acetic acid (3:1). Fixed cells were dropped on wet, warmed slides and air-dried under humidified atmosphere (Hayata etal. 1992). Slides were dried and stained with 2 % Giemsa in phosphate buffered solution (pH 6.8).

Irradiation

Exposures to carbons as well as neons with the 135 MeV/u Riken Ring Cyclotron (RRC) were performed at the Institute of Physical and Chemical Research (Riken). Exposures to carbons with the 290 MeV/u HIMAC synchrotron and those to neons with 400 MeV/u were performed at the National Institute of Radiological Sciences (NIRS) in Japan. The dosimetry and irradiation procedures were already described with details in several reports (Kanai et al. 1993). In the case of RRC, the particle fluence of ion beams was measured using plastic scintillator, depth-dose distribution with a parallel-plate ionization chamber by changing absorbers. The energy and irradiation dose were determined by multiplying the fluence by the LET with different thickness of Lucite absorbers. In HIMAC, the physical equipments were principally the same as those in RRC, although the main function was heavy ion therapy (Kanai et al. 1994, and Kanai et al. 1994). In both of the facilities, cellular materials such as cultured cells as well as blood lymphocytes were able to be irradiated with either of spread out Bragg peak (SOBP) or unspread Bragg peak. In the present study, monoenergetic beams were used for the irradiation of the blood samples. LET value of ion beams was estimated by a calculation code that took into account the fragmented nuclei. The irradiation field was defined by use of an iron and brass collimator.The depth position along irradiation path was adjusted by a Lucite range shifter. X-rays under the operation at 100 kVand 15 mA with 1.0-mm Al and 0.5-mm Cu filter were used as the reference with dose rate of 0.6 Gy/min at a FSD of 150 cm.

Chromosome Analysis

Chromosome aberrations were classified according to Nias's modification (1990) from Evans's original work (1969 and 1972) and also to IAEA classification (1986). Since the chromosomes were stained only with conventional Giemsa method, those of exchange type aberrations such as pericentric inversion and symmetrical interchanges were not distinguished. In contrast, those of asymmetric types aberrations were morphologicaly distiguishable and easy to score in metaphase plates. In this study, morphological changes due to pericentric inversion and reciprocal exchange were excluded from the aberration scoring. Chromosome analysis was forcused on the frequencies of the following 2 categories of asymmetric aberrations; 1) dicentrics and centric rings and 2) an paired linear isochromatid fragments, circular acentric rings, minutes and interstitial dots. Among these, the interstitial deletions and minutes were sometimes ambigous to differentiate.The fragments were derived from all events including terminal deletion as well as centric ring and dicentric formations. Eventually, an acentric ring was scored as a fragment. The data, based upon 100~200 aberrant first post-irradiation mitotic cells in each dose group, were used to establish the dose-aberration frequencies as a function of LET and species dependence of ion beams for aberration induction. For the sake of clarity, only the data at low dose region (1-4 Gy) were applied for analysis and were fitted to the linear-quadratic model ($Y= \alpha D + \beta D^2 + C$) to obtain dose-effect coefficients and goodness of fit for the yields of the two groups of aberrations. In order to see the effect of the particle radiations, change in frequencies of the excess fragmentation, which was defined operationally as the amount obtained by subtracting that of the categorie 1 aberrations from that of the category 2 aberrations, was described at each dose group.

RESULTS

Dose Responses for Dicentrics and Centric Rings with Different LET Values of Carbon and Neon ions.

Change in the frequencies of category 1 aberrations on per cell basis was explored as a function of dose with carbon and neon beams and a total of 7 dose responses were described in Figures 1 and 2. Carbon beams includes 3 different LET values (22.4, 41.5 and 69.9 KeV/μm) from RRC (135 MeV/u) and 2 differen LET values (40 and 100 KeV/μm) from HIMAC (290 MeV/u). Neon beams includes 70 KeV/μm from RRC (135 MeV/u) and 150 KeV/μm from HIMAC (400 MeV/u). Figure 1 shows 5 dose responses for the different LET values of carbons together with that for X-rays as a reference. Three types of dose response were seen with changes of LET. Low efficient response was found for 22.4 and 40 KeV/μm carbon beams as well as for X-rays. Medium efficiency observed for 41.5 and 100 KeV/μm carbon beams showed a linear response initially but turned into up or down concavity as the dose increased. The response of 69.9 KeV/μm of carbons revealed a linearity through the range of dose (1-4 Gy). This maximum, however, was soon cancelled by the response of 100 KeV/μm carbons, which was found to be

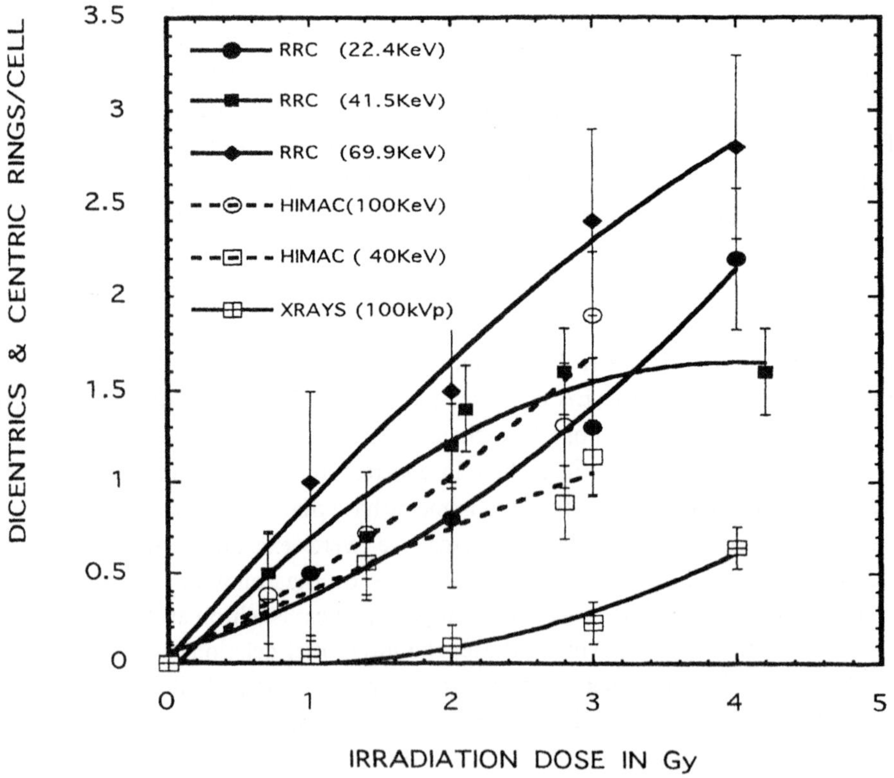

Figure 1. Dicentrics and centric rings vs. irradiation dose of carbon ions from RRC and HIMAC. The LET value of the beams is indicated by the numericals in the parenthesis in KeV/μm. The dose response with X-rays (100 kVp) is shown as the reference. Bars indicates standard errors.

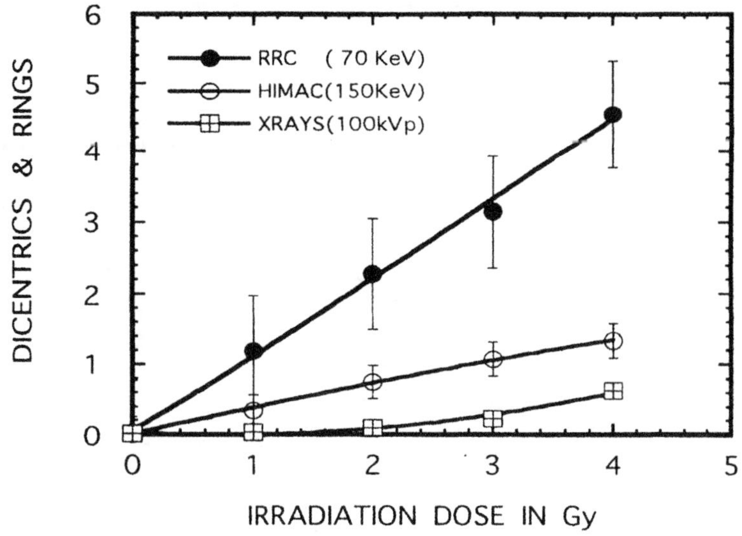

Figure 2. Dicentrics and centric rings vs. irradiation dose of neon ions from RRC and HIMAC. The LET value of the beams is indicated by thenumericals in the parenthesis in KeV/μm. The dose response with X-rays (100 kVp) is shown as the reference. Bars indicates standard errors.

close to that of 41.5 KeV/μm carbons. Two carbon beams of 40 KeV/μm from HIMAC and 41.5 KeV/μm from RRC showed a discrepancy in the response. Figure 2 shows 2 dose responses for the different LET values (70 and 150 KeV/μm) with that for X-rays. A linear response was observed for both of neon beams with different LET but not for X-rays.

Dose Response for the Excess Fragmentation of Chromosome Aberrations.

The excess fragmentation was defined as the difference in frequency of aberration between category 1 and 2 at each dose group. The data were fitted to linear-quadratic equation and the regression line was obtained for each ion beams. Figure 3 showed the excess fragmentation with 3 carbon beams produced by RRC with different LET values (22.4, 41.5 and 69.9 KeV/μm). The most efficient increase of fragmentation was found for 22.4 KeV/μm carbon ions from RRC. The other 2 different LET beams from RCC showed a medium effect for induction of the excess fragmentation. All of the 135 MeV/u carbon beams from RCC showed an excess production of fragments over that of dicentrics and centric rings. For the 40 KeV/μm carbons from HIMAC, the response showed a horizontal line going along that of X-rays. The horizontal line of response suggests an equal production of dicentrics and fragments. In contrast, the 100 KeV/μm carbons from HIMAC showed an increase of the excess fragmentation as the dose increased. The results in Figure 3 simply suggested that higher LET beams produce an excess amounts of fragments. Figure 4 shows the excess fragmentation with 2 different neon ions of 70 KeV/μm with 135 MeV/u by RRC and 150 KeV/μm with 400 MeV/u by HIMAC. As it is clearly shown in Figure 4, the excess fragmentation was much enhanced as the dose increased. For the 100 KeV/μm HIMAC neon ions, the rate of the excess fragmentation is very low when compared to that of the 70 KeV/μm RRC neon ions, which showed high efficiency in the excess fragmentation at low dose. The increase of the fragmentation showed a plateau with 70 KeV/μm neon ions.

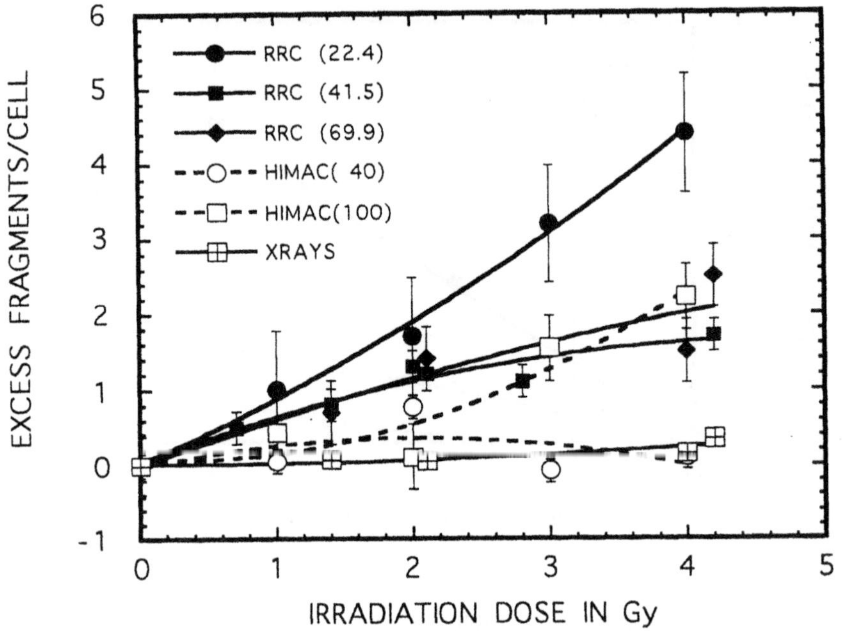

Figure 3 . The excess fragmentation vs. irradiation dose of carbon ions from RRC and HIMAC. The LET value was indicated by the numericals in the parensis in Kev/μm. The dose response with 100 kVp of X-rays is the reference. Bars indicates standard errors

Change of Relative Biological Effectiveness (RBE) with LET.

RBEs of carbons and neons for 7 different beams were calculated relative to X-rays for the coefficient α obtained by fitting the data of the group 1 and 2 to linear-quadratic equation and plotted as a function of LET in Figure 5. The RBE showed a peak around 70 KeV/μm in both of group 1 and 2 aberrations. Similar change in RBE vs. LET relation for 10 % survivals of several dsb deficient mutant and their parent cell line has been obtained with those charged particles produced by RRC (135 MeV/u) and HIMAC (290 MeV/u) with varying LET, where the mutant cells showed a peak around 80 KeV/μm and normal cells around 100 KeV/μm (Eguchi-Kasai et al. 1998). The peak position of RBE is localized close to each other in the present experiments and those by Eguchi-Kasai et al. (1998).

DISCUSSION

Measurement of aberrations in the present study was quite limited by using only the conventional Giemsa staining in contrast to other powerful techniques such as fluorescence in situ hybridization (FISH) for painting chromosomes which provide a more accurate assessment of radiation induced aberrations (Tucker et al. 1993, Savage and Simpson 1994, Edwards 1997, Testard et al. 1997). For many years, the measurements of dicentrics has been used to establish dose-response relationship in biological dosimetry. The usefulness of dicentrics has been demonstrated in the lymphocyte irradiation experiments with photons, neutrons and particle radiations of varying LET (Lloyd and Purrot 1981, IAEA 1986,

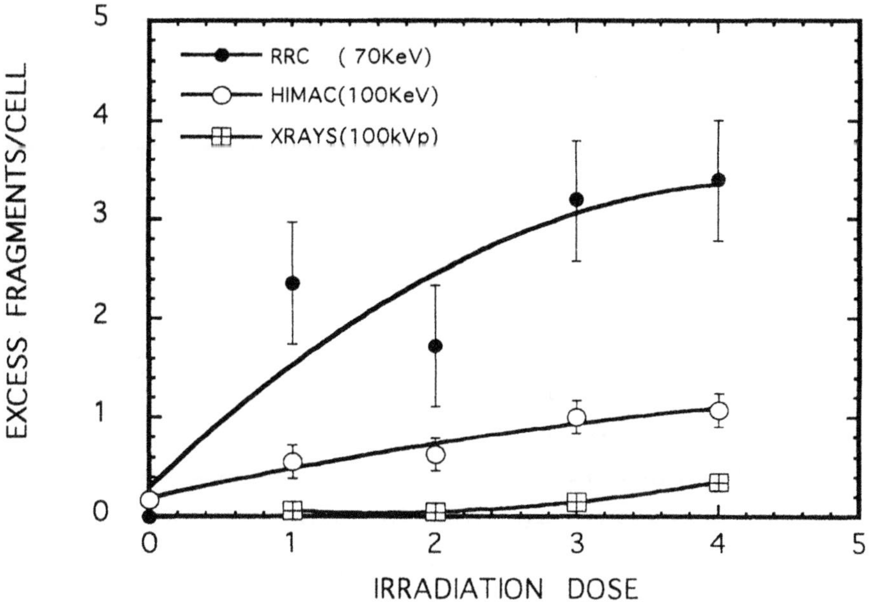

Figure 4. Excess fragments vs. irradiation dose of neon ions from RRC and HIMAC. The LET value of each ion beams is indicated by the numericals in the parenthesis in Kev/μm. The dose response of 100 kVp X-rays is the reference. Bars indicates a standard error.

Edwards 1997). In general, 2 types of response curves, a linear for high- and an upward-concave curve for low-LET radiations, were obtained by linear-quadratic regression. This may imply that the probability of two track aberrations is essentially low with low LET radiation, though it increases as LET is increased. The dose response results of Figures 1 and 2 are the representatives of chromosome damage due to 7 different carbon and neon ion beams in the present study and are similar to those previously published (Lloyd and Purrot 1981, Edwards 1997). Comparing those 7 different dose response curves, the response of carbon ions with 41.5 showed a plateaued response as irradiation dose increased. The plateaued change may arise generally due to death of cells that are severely damaged by increased chromosome aberration production. Loss of high dose experimental points may also be due to division delay induced by high LET of particle radiation (Lucke-Huhle 1979, Ritter 1996, Sholz et al., 1998). Such transitional type of curve, somehow, has not been described as a function of LET. The 2 responses of carbons and neons with the LET values of 69.9 and 70 KeV/μm (Figures 1 and 2) are considered to exhibit a complete linearity, whilethe curves of carbons and neons with LET value of 22.4 and 150 KeV/μm respectively, are upward concave. The curves of carbons with 40.0 and 41.5 showed a discrepancy in the shape of curve and in their initial slope (coefficient α) as well. The linear curves suggests that the same LET causes the same biological effect despite difference in ion species, i.e., carbons and neons with the same energy (135 MeV/u) and that the maximum effect in the present study is located around this region of LET.

The acentric fragments scored at post-irradiation metaphases may include either those from dicentric (or ring) formation or those from terminal deletion. If the production of acentric fragments is over the total of

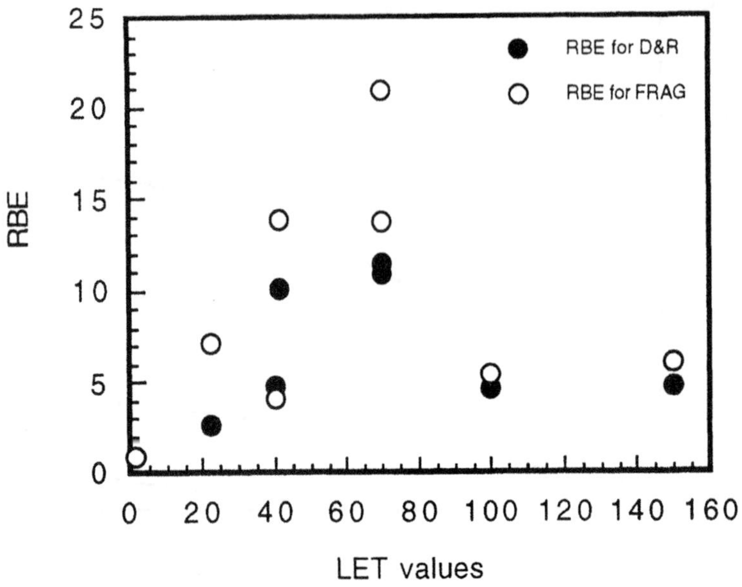

Figure 5. Change in RBE for the induction of dicentrics and centric rings (D&R) as well as of acentric fragments (FRAG) with 7 different LET values of carbon and neon ions in terms of the coefficient α relative to X-rays. The lines were obtained by general regression method.

dicentrics and centric rings production, the most of acentric fragments may be derived from unrejoining of chromosome breaks given by irradiation to result in those acentric fragments like terminal deletion or from various intra-chromosome type aberration. Thus, an excess increase of acentric fragments suggests an increase of failure (or error) of rejoining after irradiation. Figures 3 and 4 illustrate the dose effects for the excess fragmentation as defined above with carbons and neons. Thus, the horizontal response for the 40.0 KeV/μm carbons as well as X-rays (Figure 3) demonstrates that the yield of dicentrics and centric rings is almost equal to that of fragments. The results, however, on the response for 3 different carbon ions from RRC (Figure 3) indicated that the excess fragmentation is generally dependent on LET. The 22.4 KeV/μm carbon beams showed a quite high efficiency for the induction of the excess fragmentation. The fragmentation with the 100 KeV/μm carbon beams also increased as the dose increased. On the other hand, the neon beams (Figure 4) clearly showed a high efficiency in the excess fragmentation with 70 KeV/μm but not with 150 KeV/μm. This result is in contrast to those previously published studies (Testard et al. 1997, Ritter 1992, Skarsgard 1969). It has been reported that neons induce a high frequency of breaks and complex rearrangements which would not have been detected using a standard staining assay (Testard et al. 1997), and that high LET particle beams enhances the production of unrejoind chromosome breaks which may have initially originated from two unrepaired DNA dsb (Ritter, 1987 and 1996, Suzuki et al. 1997). It should be noted from the present study that high dose of neons resulted in poly-centrics, a complexed interchromosomal exchange involved at least 3-4 chromosomes, and that sometimes asymmetrical intrachanges of chromatic exchange that were not expected for Go lymphocytes were observed as well (data is not shown). In view of the results in Figures 3-4, it is likely that the excess induction of the acentric fragments depends on LET, and also that it is fundamentally derived from the increased failure in rejoining of the damaged chromosome.

The relationship between LET and induction of biological endpoints has been widely explored by many investigators in the past decades. Reproductive death and chromosome damage in mammalian cells have been shown to be more effciently produced as the LET of radiation increases, reaching a maximum in the region of 100 ~ 200 KeV/μm (Barendsen 1964, Skarsgard et al. 1967, Blakely et al. 1984 and Suzuki et al. 1996). As the most important factor of cellular lesions and death, LET dependence of DNA strand breaks and chromatin breaks have also been investigated (Ritter et al. 1997, Suzuki et al. 1996). Suzuki et al. (1996 and 1997) have shown that the RBE for cell death and the frequency of unrejoined chromatin breaks by carbons peaks around 100 -200 KeV/μm, and by neons near 100 KeV/μm. The change in RBE for the linear term coefficient α in the present study are shown in Figure 5 relative to X-rays. The plots of RBEs vs. LET indicates a peak around 70 keV/μm. Similar experimental results were obtained with several dsb deficient mutants irradiated with the particle radiations with varying LET by RRC and HIMAC (Eguchi-Kasai et al. 1998). According to Suzuki et al. (1997) and Sasaki et al. (1997), the maximum RBE for cell death and unrejoined chromatin breaks in cultured cells was found to be between 100 ~ 200 KeV/μm using the experimental conditions almost identical to the present carbon and neon experiments. The discrepancy has not yet well analyzed. More experiments are now underway for further development of the study on chromosome analysis.

ACKNOWLEDGEMENTS

This work was supported by the Co-operative Heaavy Ion Reseach Project at HIMAC in NIRS. We thank Dr. Y.Furusawa and the other staff members at the Space and Particle Radiation Reseach Group at NIRS and also at the RRC at Riken for their help in irradiation of samples and for providing information on irradiation dose and LET measurements. Thanks are also to Dr. I.Hayata and his colleagues at the Division of Radiobiology and Biodosimetry at NIRS for their help in chromosome preparation.

REFERENCES

Barendsen,G.W., Sublethal damage and DNA double strand breaks have similar RBE-LET relationships:evidence and implications. *Int. J. Radiat. Biol.* **63**, 325-330 (1993).

Barendsen, G. W., 1997. Parameters of linear-quadratic radiation dose effect relationships: dependence on LET and mechanisms of reproductive cell death., *Int. J. Radiat. Biol.*, **71**, 649-655.

Blakely, E.A. et al., Heavy-ion radiobiology : Cellular studies, *Adv. Radiat. Biol.*, **11**, 295-389 (1984).

Edward, A. A., The use of chromosomal aberations in human lymphocytes for biological dosimetry., *Radiat. Res.*, **148**, S39-S44 ,(1997)

Eguchi-Kasai, K. et al., Repair of DNA double-strand breaks and cell killing by charged particles. *Adv. Space Res.*, **22**, 543-549(1998).

Evans,H. J., Chromosome aberrations induced by ionizing radiations, *Int. Rev. Cytol.*, **13**, 221-321 (1962).

Evans, H. J., Ations of radiations on human chromosomes. *Physics in Medicine and Biology.*, **17**, 1-13 (1972).

Furusawa, Y., K. Fukutsu, H. Itsukaichi, K. Eguchi-Kasai, H. Ohara, F. Yatagai, and T. Kanai, Biological effectiveness by the different heavy ion species with the LET. In : Proceedings of the Second Workshop on Physical and Biological Research with Heavy Ions, Edited by K. Ando and T. Kanai., *NIRS Report NIRS-M-***90**, HIMAC-003 (Chiba), pp. 11-13 (1992).

Hayata, I., H. Tabuchi, A. Furukawa, N. Okabe, M.Yamamoto, and K. Sato, Robot system for preparing lymphocyte chromosome. *J. Radiat. Res.*, **33** (suppl.), 231-241 (1992).

IAEA *Technical Reports* No.**260**, Biological dosimetry: Chromosomal aberration analysis for dose asessment. IAEA, Vienna.(1986).

Kanai , T., S. Minohara, T. Kohno, M. Sudou, E.Takada, F. Soga, M. Kawach, A. Fukumura., and F. Yatagai, Irradiation of 135 MeV/u carbon and neon beams for studies of radiation biology. *NIRS Report NIRS-M*-**103**, HIMAC-008, (Chiba) 1-39 (1993).

Kanai T., H.Tomura, N. Mtsufuji, N. Minohara, H. Koyama-Itoh, M. Endo, and F. Soga, HIMAC beam delivery system Physical characteristics. *Proceedings of NIRS international seminar on the application of heavy ion accelerator in connection with XXI PTCOG meeting, NIRS-M*-**103**, HIMAC-008, (Chiba:NIRS) 26-31 (1994).

Kraft, G. , Radiobiological effects of very heavy ions: Inactivation, induction of chromosome aberrations and strand breaks. *Nuclear Science Application*, **3**, 1-28 (1987).

Lloyd, D. C., and R.J. Ourrott, Chromosome aberration analysis in radiological protection dosimety, *Radiation Protection Dosimetry*, **1**, 19-27 (1981).

Lucas, J. N., A.M. Chen, and R.K. Sachs, Theoretical predictions on the equality of radiation-prodeced dicentrics and translocations detected by chromosome painting. *Int. J. Radiat. Biol.*, **69**, 145-153 (1996).

Lucke-Hohle, C., E.A. Blakely, P.Y. Chang, and C.A. Tobias, Drastic G2 arrest in mmamalian cells after irradiation with heavy ion beams. *Radiat. Res.*, **79**, 97-112.

Nias, A. H. W., *Chromosome Damage in An introduction to Radiobiology*, John Wiley & Sons, Chichester, pp. 67-75 (1990).

Ritter, S., W. Kraft-weyrather, M. Scholz, and G. Kraft, Induction of chromosome aberrations in mammalian cells after heavy ion exposure. *Adv. Space Res.*, **12**, 119-125 (1992).

Ritter, S., Comparison o chromosomal damage induced by X-rays and Ar ions with LET of 1840 KeV/μm in G1 V79 cells. *Int. J. Radiat. Biol.*, **69**, 155-166 (1996).

Ritter, M.A., J. E. Cleaver, and C. A. Tobias, High-LET radiations induce a large proportion of non-rejoining DNA breaks. *Nature*, **266**, 653-655 (1997).

Testard, I., B. Dutrillaux, and L. Sabatier, Chromosomal aberrations induced in human lymphocytes by high LET radiation., Int. *J. Radiat. Biol.*, **72**, 423-433 (1997).

Tucker, J.D., M.J. Ramsey, D.A. Lee and J. L. Minker, Variation of chromosome painting as a biodosimeter in human peripheral lymphocytes following acute exposure to ionizing radiation in vitro. *Int. J. Radiat. Biol.*, **64**, 27-37 (1993).

Savage, J. R. K. and P. J. Simpson , FISH"painting" patterns resulting from comples exchanges, *Mutation Res.,* **312**, 50-60 (1994).

Sasaki, H. , F. Yatagai, T. Kanai, Y. Furusawa, F. Hnaoka , Wei-Guo Zhu and P. Mehnati, Dependence of induction of interphase death of Chinese hamster Ovary cells exposed to accelerated heavy ions on linear energy transfer, *Radiat. Res.*, **148**. 449-454 (1997).

Sholz, M., S. Ritter, and G. Kraft, Analysis of chromosome damage based on the time course of aberrations, Int. J. Radiat. Biol., **74,** 325-331(1998).

Skarsgard, L.D., Kihlman, B. A., Parker, L., Pajara, C. M. and Richardson, S., Survival, chromosome abnormalites, and recovery in heavy ions and X-rays. *Radiat. Res.*, **7** , (Suppl.), 208-221. 1967.

Pergamon

Adv. Space Res. Vol. 22, No. 12, pp. 1683–1690, 1998

Printed in Great Britain
0273-1177/98 $19.00 + 0.00

PII: S0273–1177(99)00033–2

CYTOGENETIC EFFECTS OF ENERGETIC IONS WITH SHIELDING

T. C. Yang[1], K. A. George[1,2], H. Wu[1,3], D. Miller[4], and J. Miller[5]

[1]NASA Johnson Space Center, Radiation Biophysics Laboratory, Houston, TX 77058, USA
[2]Wyle Laboratories, Houston, TX 77058, USA
[3]Kelsey-Seybold Clinic, Houston, TX 77058, USA
[4]Loma Linda University Medical Center, Loma Linda, CA 92350, USA
[5]Lawrence Berkeley National Laboratory, University of California, Berkeley, CA 94720, USA

ABSTRACT

In order to understand the effects of shielding on the induction of biological damages by charged particles, we conducted experiments with accelerated protons (250 MeV) and iron particles (1 GeV/u). Human lymphocytes *in vitro* were exposed to particle beams through polyethylene with various thickness, and chromosomal aberrations were determined using FISH technique. Dose response curves for chromosome aberrations were obtained and compared for various particle types. Experimental results indicated that for a given absorbed dose at the cell, the effectiveness of protons and iron particles in the induction of chromosomal aberrations was not significantly altered by polyethylene with thickness up to 30-cm and 15-cm respectively. Comparing with gamma rays, charged particles were very effective in producing complex chromosomal damages, which may be an important mechanism in alterating functions in non-dividing tissues, such as nervous systems.

INTRODUCTION

One of the major problems in estimating space radiation's health effects for astronauts is the uncertainty of the actual particle distribution to crew members inside a spacecraft, inside a space suit when conducting extravehicular activities, or at specific organs of crew members (National Research Council Report, 1996). When space radiation penetrates through the shielding material of space craft or human body, protons and heavy ions will produce secondary particles. These secondary particles can have lower energies with lower or higher mass and can be more effective in causing biological effects than primary ions (Wilson et al., 1997; Yang and Craise, 1997). For health risk assessment, it is thus necessary to determine the potential effects of various shielding materials on the effectiveness of charged particles in producing important biological damages in human cells. In order to understand the shielding effects, we conducted experiments with accelerated protons and iron particles. Human lymphocytes were irradiated with particle beams through shielding material of various thickness, and the frequency of chromosomal aberrations were determined. Results from these experiments are analyzed, compared with that of gamma rays, and reported here.

MATERIALS AND METHODS

Techniques and methods for studying chromosomal aberrations in human lymphocytes were reported in detail before (Yang et al., 1997; Wu et al., 1998). Briefly, whole human blood samples were collected in plastic tubes containing sodium heparin (100 USP units) and after irradiation, were incubated in culture medium [Gibco RPMI 1640 medium supplemented with 20% calf serum and 1% phytohemagglutinin (PHA)]. Colcemide (0.2 mg/ml) was added after 48 hr incubation at 37°C. The cells were then harvested about 2 hr after Colcemide treatment. After the cells were swollen in 0.075 M potassium chloride, they were fixed in methanol/acetic acid (3:1) and stored at -20°C. Chromosome spreads were prepared by dropping cells onto clean glass slides. Chromosome spreads were then hybridized with fluorescence-labeled DNA probes specific for chromosome 2, 3 or 4. All other chromosomes were counter-stained with 4,6-diamidino-2-phenylindole (DAPI). Chromosomes were viewed with a Zeiss Axioplan fluorescence microscope. Images of aberrant chromosomes were captured by computer and further enhanced using the Power Gene Analysis System from Perceptive Scientific Instruments (League City, Texas). A pair of biocolor chromosomes were scored as a reciprocal exchange. The frequency of incomplete exchanges, complex exchanges and excess fragments was also noted. Aberrations were recorded using the PAINT classification system (Tucker et al., 1995). A sub-class of complex-type exchanges involving one painted chromosome was considered. This category excluded the non-reciprocal complex exchanges (3 chromosomes involved but only one break/chromosome).

IRRADIATION

Freshly collected whole blood samples in a 15-ml plastic tubes with a diameter of 1.6 cm were irradiated at room temperature with gamma rays, protons or iron particles. Gamma irradiation utilized a ^{137}Cs source at an absorbed dose rate of 10 Gy/min at Baylor College of Medicine, Texas Medical Center, Houston. Proton beams were provided by the proton accelerator at the Loma Linda University Medical Center, Loma Linda, California. The beam energy of the protons was 250 MeV, corresponding to a track average LET in tissue of about 0.4 keV/μm, and the dose rate was about 0.3 Gy/min. Iron beam irradiation was performed at the Brookhaven National Laboratory, Upton, New York. Detailed characterization of the 1087 MeV/u iron beam has been reported by Zeitlin et al., (1998). The track average LET in tissue of iron particles was about 140 keV/μm, and the dose rate was between 0.5 and 1 Gy/min. For shielding studies, polyethylene with various thickness was placed between the in-coming particle beam and the cell sample. Dose calibration for samples irradiated with polyethylene was calculated using EG&G instrument.

RESULTS

Table 1 shows the experiment data for lymphocytes exposed to protons with or without polyethylene shielding. The residual range of proton beam was measured at about 0.3-cm and 31.5-cm in water with and without polyethylene, respectively. Within the 1 to 7 Gy doses used, no statistical difference was found between cells exposed with or without polyethylene with respect to either percent of aberrant spreads or percent of reciprocal exchanges.

The percent of aberrant spreads found in cells exposed to protons without shielding as a function of doses is given in Figure 1. The dose response curve is curvilinear with initial slope of 0.88/Gy. Figure 2 shows the comparison between cells irradiated with or without polyethylene for reciprocal exchange frequency. Clearly there was no significant difference. One interesting effect, however, was observed.

Table 1. Chromosomal aberrations (in chromosome numbers 2 and 4) induced by protons with or without polyethylene shielding.

Shielding Thickness	Dose (Gy)	No. Cells Scored	No. Aberrant Spreads [per cell]		No. Reciprocal Exchanges [per cell]	
0.0 cm	0	676	2	[0.0029]	2	[0.0029]
	1	1198	33	[0.0275]	8	[0.0067]
	3	286	34	[0.1189]	18	[0.0629]
	5	143	50	[0.3496]	31	[0.2168]
	7	64	40	[0.6250]	27	[0.4219]
31.2 cm	1	454	21	[0.0462]	10	[0.0220]
	3	178	38	[0.2135]	20	[0.1124]
	5	150	64	[0.4267]	48	[0.3200]
	7	96	69	[0.7185]	42	[0.4375]

There were cells showing multiple breaks in one chromosome while the other homologous one was apparently normal. Table 2 shows the data for those single chromosomes with 3 or more breaks. The dose was calibrated at the sample, and chromosome #2 and #4 were painted for analyses. With polyethylene shielding, the LET increased from about 0.4 keV/μm to about 2.6 keV/μm as the energy decreased from about 235 MeV to about 20 MeV. Proton beam with shielding appeared to cause more multiple chromosomal damages in lymphocytes. Due to small sample size, the difference was not statistically significant. However, for a given dose, there were consistently more cells which showed multiple chromosomal breaks with shielding.

Experimental results of lymphocytes exposed to 1 GeV/u iron particles are given in Table 3, Fig. 3 and 4. Figure 3 shows the dose response curves for total exchanges per cell as a function of dose. There was no significant difference between 0-, 5-, and 15-cm polyethylene shielding, and the dose response curves were non-linear. All data points, for a given dose, were very close together, except the data point of 4 Gy for 0-cm shielding. The data given in Table 3 were from FISH analyses of aberrations in chromosome #2. When the comparison was made for frequency of aberrant spreads, there was no significant

Table 2. Proton Induced Multiple Chromosomal Damages (chromosomes 2 and 4) in Human Lymphocytes.

Residual Range in Tissue	Energy at Sample (MeV)	LET (keV/ μm)	Dose (Gy)	# Cells with Multiple Damages	Total # Cells Scored	% Cells with Multiple Damages + SD
31.5 cm	235	0.4	0.00	0	633	0.00 ± 0.0
			1.50	1	471	0.21 + 0.21
			2.25	1	395	0.25 + 0.25
			3.00	1	329	0.30 + 0.30
0.30 cm	20	2.6	1.50	2	641	0.31 + 0.22
			2.25	2	247	0.81 + 0.57
			3.00	2	229	0.87 + 0.61
			3.75	3	186	1.61 + 0.93

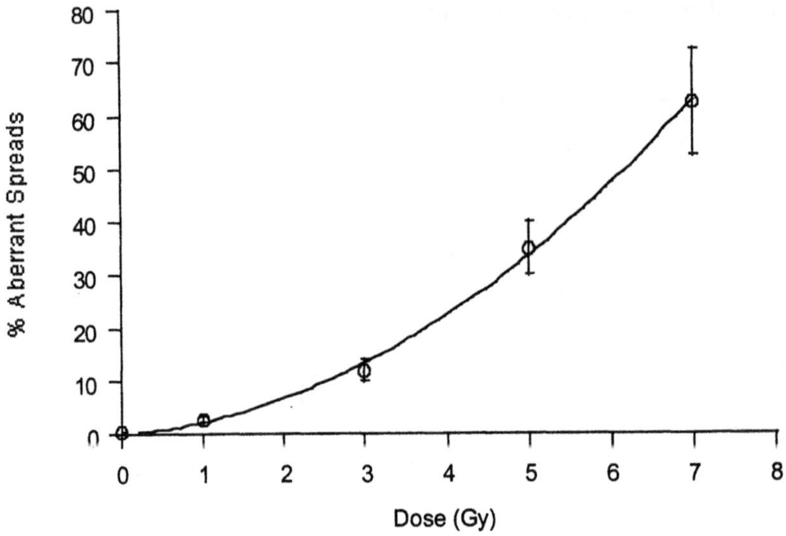

Figure 1. The percent of spreads with aberrant chromosome 2 or 4 found in human lymphocytes exposed to 250 MeV protons as a function of dose.

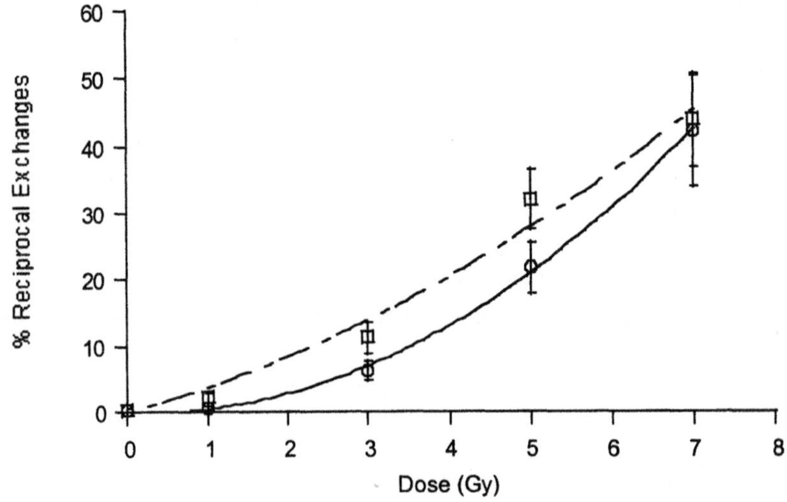

Figure 2. A comparison between lymphocytes irradiated with (open square) or without (open circle) polyethylene for reciprocal exchanges found in chromosomes 2 and 4. The residual range of protons with shielding was 0.3 cm in tissue.

difference between 0-, 5- and 15-cm shielding also. Polyethylene shielding up to 15-cm in thickness, therefore, did not alter the effectiveness of 1 GeV/u iron beam in causing chromosome aberrations in human lymphocytes.

Table 3. Effects of Polyethylene Shielding on the Induction of Chromosomal Aberrations (chromosome 2) in Human Lymphocytes by 1 GeV/u Iron Particles.

Dose (Gy)	Shielding Thickness	Cells Scored	No. Aberrant Spreads [per cell]	Total Visible Exchanges [per cell]
0	0 cm	416	1 [0.0024]	2 [0.0048]
0.5		437	19 [0.0435]	24 [0.0549]
1.0		422	38 [0.0900]	59 [0.1398]
2.0		254	37 [0.1457]	65 [0.2559]
3.0		189	39 [0.2063]	73 [0.3862]
4.0		255	40 [0.1569]	84 [0.3294]
0.5	5 cm	451	19 [0.0421]	37 [0.0820]
1.0		373	19 [0.0509]	41 [0.1099]
2.0		102	20 [0.1961]	27 [0.2647]
3.0		198	37 [0.1869]	78 [0.3939]
4.0		67	25 [0.3731]	46 [0.6866]
0.5	15 cm	467	12 [0.0257]	22 [0.0471]
1.0		424	21 [0.0495]	36 [0.0849]
2.0		197	29 [0.1472]	44 [0.2233]
3.0		157	39 [0.2484]	67 [0.4267]
4.0		50	19 [0.3800]	31 [0.6200]

Table 4. Induction of Complex Exchanges and Multiple Damage in a Single Chromosome by Gamma Rays or 1 GeV/u Iron Particles. (0 cm shielding)

Radiation Type	Chromosome Analyzed	Dose (Gy)	Frequency of Complex Exchanges	Frequency of Multiple Damages in One Chromosome
Gamma	#3	0	0	0
Rays		1	0.001 ± 0.001	0.001 ± 0.001
		2	0.008 ± 0.004	0.003 ± 0.003
		3	0.008 ± 0.004	0.004 ± 0.003
		4	0.02 ± 0.009	0.01 ± 0.007
		5	0.05 ± 0.02	0.03 ± 0.01
Iron	#2	0	0	0
Particles		0.5	0.01 ± 0.005	0.007 ± 0.004
		1	0.03 ± 0.008	0.016 ± 0.006
		2	0.06 ± 0.02	0.024 ± 0.01
		3	0.09 ± 0.02	0.032 ± 0.01
		4	0.09 ± 0.02	0.051 ± 0.01

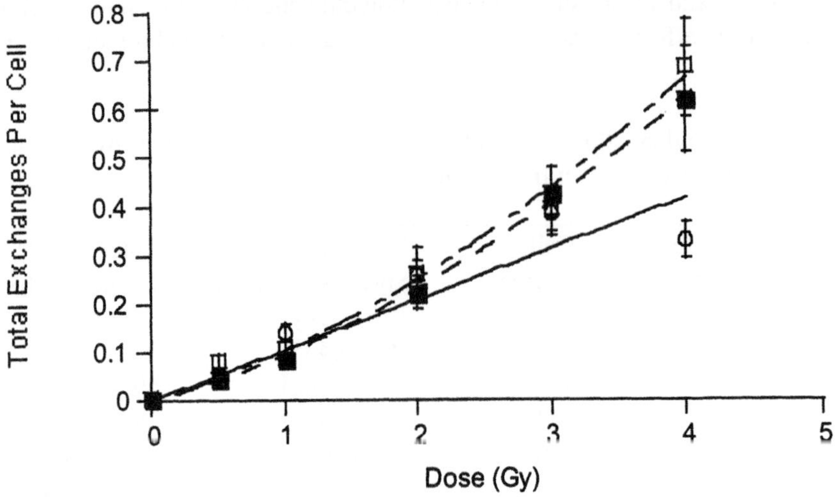

Figure 3. Dose-response curves for total exchanges in chromosome 2 per cell as a function of dose. Lymphocyte were irradiated with 1 GeV/u iron particles behind 0- (open circle), 5- (open square) or 15-cm (solid square) polyethylene.

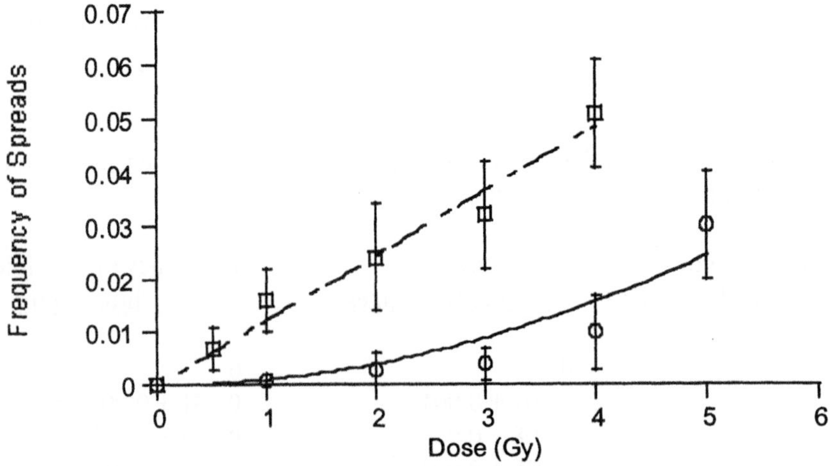

Figure 4. A comparison between 1 GeV/u iron particles and gamma rays in inducing cells with one normal chromosome and one chromosome with two or more breaks.

A comparison between iron particles and gamma rays in inducing lymphocytes with one normal chromosome and one chromosome with two or more breaks is shown in Table 4 and Fig. 4. Iron particles evidently were much more effective in causing multiple breaks in chromosomes.

DISCUSSION

Space radiation consists of energetic charged particles, including protons and heavy ions, which pose potential health problems to space travelers (Schimmerling, 1992). These charged particles have sufficient energies to penetrate through spacecraft and human body. For radiation protection of astronauts, it is legally required to assess health risks before and after each mission. Accurate risk assessment requires not only accurate physical dose measurements but also accurate organ dose calculation. For organ dose estimation, transport codes for charged particles and computerized human anatomic models have been used. While these transport codes and anatomic models are valuable for health risk assessment, they are not without significant uncertainty. Although many studies were performed to determine the effectiveness of charged particles in causing biological damages, very few experiments have been designed and conducted to determine the effects of shielding (Yang and Craise, 1997). In view of the needs of experimental data on shielding effects, we performed experiments with 250 MeV protons and 1 GeV/u iron particles. These charged particles and energies were chosen for practical reasons. Protons with energies up to several hundreds MeV contribute more than half of the absorbed dose for Shuttle and Mir flights, and energetic iron particles are heavy ions found in cosmic rays and highly effective in causing biological damages. Polyethylene was used as shielding material for its density is close to human tissue. In space, human body provides a major shielding of radiation.

Our experimental results showed that for a given dose at the sample, polyethylene up to 30 cm and 15 cm had no effect on the induction of chromosomal aberrations in human lymphocytes by 250 MeV protons and 1 GeV/u iron particles, respectively. These are interesting results, particularly data for the iron particles. Physical measurements with 1 GeV/u iron beam was performed and reported by Zeitline et al. (1998). According to these measurements, the track-average LET decreased with increasing of thickness of polyethylene, as shown in Table 5. For 1 GeV/u iron beam, the LET changed from about 147 keV/μm to about 95 keV/μm as the thickness polyethylene increased to 9.35 cm. The mean free path for fragmentation of iron particles in water is 10 cm and at 12 cm only about 30% of the incident iron particles survive. With further increase in thickness to 29.7 cm, only about 5% of the incident iron particle survive. For iron beam experiments, we selected 0-cm, 5-cm and 15-cm polyethylene for the interest of responses in skin, blood forming organ and tissues in mid of body. After passing through 15-cm polyethylene, there should be less than 30% of primary iron particles survived. Yet, no change of chromosomal aberration frequency was observed in cells shielded with 15-cm polyethylene. The

Table 5. Track-Average LETs of Iron Beam Behind Polyethylene

Energy (MeV/u)	Polyethylene (g/cm^2)	LET (keV/μm)
1087	0.00	147
	1.94	135
	4.68	117
	9.35	95

secondary ions produced by fragmentation of target somehow compensated the loss of primary iron particles in terms of inducing chromosomal aberrations. In the future, experiments should be done with shielding greater than 15 cm, e.g., 20 and 25 cm, to determine if the present results continue to hold.

One of the unexpected findings of proton experiments was the induction of multiple damages in only one of the two homologous chromosomes. The consistent increase of frequency of such damages in cells with thick polyethylene shielding suggests that this might be due to secondary particles, including neutrons. The low frequency of such damage found even in cells exposed to high doses supports this notion. Unfortunately, the low frequency of aberrations also makes the statistics of data undesirable and thus a final conclusion of this observation difficult. It is necessary to conduct more experiments with protons in the future. These proton experiments should have both physical measurements of secondary particles and analyses of multiples damages in chromosomes.

Interestingly, multiple breaks in one of the two homologous chromosomes were also observed in lymphocytes exposed to 1 GeV/u iron beam. This indicates that high density of ionization in iron particle track is important for such severe effects. The effectiveness of heavy ions in causing multiple breaks in one chromosome may related to the spatial distribution of ionization as well as the chromosomal domain in the cell nucleus. Perhaps, biophysical model can be developed in the future to account for this type of damages. This type of damages may have great importance in causing functional alterations in non-dividing cells, such as neurons in the brain. Multiple breaks of one chromosome may lead to permanent loss of the function of this chromosome. The cell will then become more vulnerable to additional insults in terms of loss specific functions, since there is only one copy of gene functioning in the intact chromosome. Further studies of this type of damages in neurons and central nervous system are warranted.

REFERENCES

National Research Council Report, *Radiation Hazards to Crew of Interplanetary Missions: Biological Issues and Research Strategies*, National Academy Press, Washington, D.C. (1996)

Schimmerling, W., Radiobiological Problems in Space: An Overview. *Radiat. Environ. Biophysic.*, **31**, 197-203 (1992)

Tucker, J. D., W. F. Morgan, A. A. Awa, M. Bauchinger, D. Blakely, M. N. Cornforth, L. G. Littlefield, A. T. Natarajan, and C. Shasserre, A Proposed System for Scoring Structural Aberrations Detected by Chromosome Painting, *Cytogenet. Cell Genet.* **68**, 211-221 (1995)

Wilson, J. W., F. A. Cucinotta, S. A. Thibeault, M. Kim, J. L. Shinn, and F. F. Badavi, Radiation Shielding Design Issues, in *Shielding Strategies for Human Space Exploration*, NASA Conference Publication 3360, edited by J. W. Wilson, J. Miller, A. Konradi, and F. A. Cucinotta, pp. 109-150 (1997)

Wu, H., K. George, and T. C. Yang, Estimate of True Incomplete Exchanges Using Fluorescence *in situ* Hybridization with Telomere Probes, *Int. J. Radiat. Biol.* **5**, 521-527 (1998)

Yang, T. C. and L. M. Craise, Biological Responses to Heavy Ion Exposure, in *Shielding Strategies for Human Space Exploration*, NASA Conference Publication 3360, edited by J. W. Wilson, J. Miller, A. Konradi, and F. A. Cucinotta, pp. 91-108 (1997)

Yang, T. C. , K. George, A. S. Johnson, M. Durante, and B. S. Fedorenko, Biodosimetry Results from Space Flight Mir-18, *Radiat. Res.* **148**, S17-S23 (1997)

Zeitlin, C., L. Heilbronn, and J. Miller, Detailed Characterization of the 1087 MeV/nucleon Iron-56 Beam Used for Radiobiology at the Alternating Gradient Synchroton, *Radiat. Res.* **149**, 560-569 (1998)

Pergamon

Adv. Space Res. Vol. 22, No. 12, pp. 1691–1697, 1998

Printed in Great Britain
0273-1177/98 $19.00 + 0.00

PII: S0273-1177(99)00034-4

MORPHOLOGICAL AND MOLECULAR CHANGES OF MAIZE PLANTS AFTER SEEDS BEEN FLOWN ON RECOVERABLE SATELLITE

M. Mei[1], Y. Qiu[1], Y. Sun[2], R. Huang[3], J. Yao[1], Q. Zhang[1], M. Hong[1] & J. Ye[1]

[1] *South China Agricultural University, Guangzhou 510642, China*
[2] *Harbin Institute of Technology, Harbin 150001, China*
[3] *Instate of High Energy Physics, Academic Sinica, Beijing 100039, China*

ABSTRACT

Dry seeds of *Zea mays*, heterozygous for Lw_1/lw_1 alleles, sandwiched between nuclear track detectors aboard Chinese satellite for 15 days, were recovered and mutations in morphological characters on plants developed from these seeds, as well as their selected progenies, were investigated. The dosimetric results indicated that 85% of the seeds received at least 1 hit with $Z \geq 20$. About 10% of plants developed from flown seeds and 40% of observed selfed lines from the first generation plants showed some morphological changes, such as yellow stripes displayed on leaves, dwarf, anomogensis of floral organs and yellow-green seedlings, when compared with those from ground control. Using yellow stripes on leaves as the main endpoint for evaluating mutation induced in space environment, the frequency of stripe occurrence was 4.6% in the first generation plants, comparable with the results obtained from Long Duration Exposure Facility (LDEF) mission (Mei et al., 1994), but much lower than those from ground based ^{60}Co-gamma treatment at a dose of 100 Gy, which reached 35.5% in the selfed lines of the second generation. One hundred and ten random primers were screened in RAPD analysis to detect the variation on genomic DNA of plants with stripes on leaves. Of these primers, 10.9% were able to generate polymorphic bands between mutated plants and control, also, common band patterns in several progenies with the same mutation phenotype were observed. These results demonstrated that space radiation environment could induce inheritable mutagenic effects on plant seeds, and verified the change in genetic material in the mutants. Further study will be needed for a better understand of the nature and mechanism of this induction of mutation.

INTRODUCTION

Since the beginning of human space exploration, a series of experiments on the biological effects of space radiation had been conducted in different missions. A number of reports documented that exposure of various biological systems, i.e. virus, bacterial spores, plant seeds, insect eggs etc., to space radiation, especially to the high LET cosmic ray particles, would have serious biological effects, such as developmental abnormalities, chromosome aberration, cell inactivation, somatic mutation and tumor induction. However, most of these reports presented the results from the short-term observation of the living systems, e.g. one generation, and very few related to the long-term effects of space radiation, as summarized by Horneck (1992).

Recently, the reports from China indicated that some stable mutants with promising traits, from the progenies of space flown crop seeds, such as rice, wheat, had been selected and used in the breeding of new cultivars (Jiang, 1996). To understand the mechanism of mutation induction as a result of space flight, we investigated the mutagenic effects of space flight on maize seeds; i.e. the morphological and molecular changes in the first and second-generation plants, and the results were presented here.

METHODOLOGY

Plant Materials

Dry seeds of *Zea mays,* heterozygous for Lw_1/lw_1 alleles, were obtained from Ratio Seeds Company (Indiana, USA). Mutation of Lw_1 in embryonic cell for leaf development will cause white/yellow stripe formation on leaves (Qiu et al.,1991). A total of two hundred seeds were placed on the Chinese recoverable satellite during the space flight from October 5 to October 20, 1996 at an orbit of 63°, 175 km/320 km. Among them, one hundred seeds were fixed on two layers of plastic holders in size 8 cm x 7 cm, sandwiched between the nuclear track detectors to localize the penetrating sites of HZE particles. Part of the seeds from the same stock were kept on the ground in similar environmental conditions as that of flown seeds, and one hundred seeds from this stock were irradiated with ^{60}Co gamma rays at a dose of 100 Gy as ground-based radiation control. The remaining seeds were used as control.

Field Observation

After retrieval, space flown seeds and control seeds were planted in the fields of the experimental farm at the South China Agricultural University, Guangzhou, China in March 1997. The growth, development and variation in morphological characters of resultant plants from these seeds were studied. The progenies from some of the self- bred plants from the first generation were also investigated to identify mutants in the second generation. The occurrence of white/yellow stripes on leaves was used as the main endpoint in selecting mutants.

RAPD Analysis

For understanding the molecular nature of mutagenesis induced by space flight, some of the plants with morphological change were chosen for Random Amplified Polymorphic DNA (RAPD) analysis in comparison with the control plants. Maize DNA was extracted from fresh-frozen leaf tissue with the CTAB method (Murray and Thompson, 1980) with minor modification. The PCRs for RAPD analysis were performed in 25 μl volume containing reaction buffer, 2.0 mM $MgCl_2$, 100 μM of each deoxynuclotide, 1 U of Taq DNA polymerase, 0.2 μM of 10-mers random primers from Operon Technologies (California, USA), and 20 ng of maize DNA, using thermal cycler (Perkin-Elmer Cetus, USA). The profile consisted of 30 seconds denaturation at 94°C, 1 min annealing at 37°C, 2 min extension at 72°C for 38 cycles, and finally extension for 7 min at 72°C. The amplification products were loaded into a 1.5 % (w/v) agarose gel for electrophoresis in 1x TBE buffer. Gels were stained with ethidium bromide, observed and photographed under UV light (Williams et al., 1990)

RESULTS AND DISCUSSION

Dosimetric Measurement

To investigate the biological effects of individual heavy ions to maize seeds, sandwich-like combination of seeds in monolayer fixed in between nuclear track detector, similar to that of the Biostack concept (Bucker et al., 1974), was used for localizing the trajectory of HZE particles relative to the seed. The results from alkali etching of track detectors indicated that there were 79 and 94 particles of $Z/\beta \geq 50$ ($Z \geq 20$), which traversed respectively, seed layers 1 and 2. Of these numbers, 24 embryos were hit in layer 1 and 23 in layer 2. In addition, 35% of seeds received only one hit, 50% of seeds received two hits at least, and the other 15% did not receive any hit from these types of particles. When particles of $Z/\beta > 10$ ($Z > 3$) were measured, each maize seeds received 19 hits on average, and the numbers of $Z > 3$ particles reached 29.7 ± 0.5 cm^{-2}.

Morphological Change in the First and Second Generation of Maize Plants

Among the two hundred seeds exposed to space radiation, one hundred and ninety five were able to germinate normally and grow into plants. Several types of mutation on morphological characters such as occurrence of yellow stripes on some of leaves, shortening of plant height, and flower anormogensis, i.e. absence of staminate flower, pistillate flower and visible seed in the tassel, color of seedling from normal green to yellow-green etc., were observed in the plants developed from these seeds. The types and frequencies of mutation were summarized on Table 1.

The maize seeds used in the present study contained a recessive lemon white gene, and the white/yellow stripes had been observed in leaves of plants from seeds of the same genetic stock as those used in this study, flown on the missions ASTP (Peterson et al., 1977), LEDF (Mei et al., 1994), and a couple of studies on ground radiation effects (Yang and Tobias, 1979; Qiu et al., 1991). The occurrence of large white/yellow stripes on leaves in high percentage of plants developed from seeds irradiated with ^{60}Co gamma rays, as shown on Table 1, demonstrated that the formation of stripes was a useful endpoint for evaluating somatic mutations caused by ionizing radiation. The results shown in Table 1 also indicated that the frequency (F) of stripe occurrence, i.e. the ratio of the number of plants with striped leaf to the number of plants observed, reached 4.6 %, a little lower than those flown on LDEF (5.4%). However, the latter underwent 69 months space missions, exposed to a dose of 635 cGy which was estimated from the value of white/yellow striped leaves frequency when compared with data from ground based studies (Mei et al., 1994), and much higher than the dose the seeds were exposed to in this report, estimated to be 2.65mGy by integrating dosimeter (TLD). The rather high efficiency for the induction of yellow stripes on leaves after space flight shown in this study demonstrated that a few hits of high LET charged particles from space radiation environment would be enough to induce this type of somatic mutations.

Table 1 Mutation Observed in the First and Second Generation Plants Developed from Space Flown or Gamma Irradiated Seeds

Treatments (generations)	No. Plants (lines) observed	Mutation observed: Yellow stripe occurrence		Shorten*** plant		Anormogensis of flower		Yellow-green seedlings	
		Plants (lines)	F** (%)	Plants	F** (%)	Plants	F** (%)	Plants	F** (%)
Flown (M_1)	195 (plants)	9	4.5	6	3.07	3	1.5	4	2.05
Flown (M_2)	45 (lines)*	16	35.5	18	40.0	3	6.6	0	0
^{60}Co-γ-ray (M_1)	99 (plants)	40	40.4	2	2.0	0	0	0	0
Control (M_1/M_2)	50/30	0/0	0/0	0/0	0/0	0/0	0/0	0/0	0/0

*Each line consisted of at least 30 selfed progenies of each plant developed from the flown maize seeds, and the mutations were observed in some of these progenies in each line.
**F (frequency) = (No. Plants with mutated character) / No. Observed plants)
***Plant height< 75% of the average plant height of the control group

Yellow stripes were also found on the leaves of some of the progenies in the selfed lines produced by four plants developed from seeds flown on the satellite and with yellow striped leaves. This phenotype was found in more than 10% of the selfed progenies of plant developed from seed 21, which was estimated to have received 3 particles with $Z \geq 20$ and 42 particles with $Z \geq 3$. On the other hand, expression of this phenotype in twelve selfed lines in which the first generation showed no such phenotype was found. The ratio of the number of lines in which striped leaves were expressed to the number of lines observed was as high as 35.5% (Table 1). This indicated that the mutation frequency in the second generation was rather high. A good example was the plant developed from seed number 2, in which the embryo received one particle with $Z \geq 20$, was found to be normal. However, large yellow stripes displayed on leaves of 14 selfed progenies from this plant. Another example was seed number 43, which received two hits with $Z \geq 20$ inside and outside of the embryo. The yellow striped leaves were also found in the selfed progenies of the second and third generations developed from this seed. These results suggested that the formation of yellow stripes on leaves was a result of inheritable changes induced by space radiation, and this change might exist in the subsequent generations.

A high frequency of mutation in the height of plant, i.e. shorten plant, was observed, especially in the second generation, as shown in Table 1. The results suggest that radiation in space induced similar types of mutations on plant development as on ground, since inducing of dwarf mutation is a common phenomenon for ionizing radiation. The dwarf plants selected from M_1 were sterile, but seeds from several semi-dwarf plants in one M_2 line have been obtained recently, and will be planted later this year for further observation. Other mutagenic studies performed by our colleagues in the South China Agricultural University on rice seeds from the same space flight has obtained several semi-dwarf mutants (data not showed) with stable inherited change for at least two generations.

Mutation was also observed in anormogensis of floral organ. In maize, the pistillate (ear) and staminate (tassel) inflorescence should grow separately, with staminate tassel in the main shoot terminal and pistillate ear axillary. Two types of variation among the two generations of plants developed from flown seeds were found, i.e. plants with normal pistillate, but without staminate flower; and plants produced functional pistillate flower in the tassel, and then after pollination, formed viable seeds in the tassel. This conversion of staminate to pistillate in tassel might be due to the failure of the normal pistil abortion process, and related to the function of *Ts1* and *Ts2* genes, as described by Dellaporta et al. (1994).

Molecular Changes of Maize Mutants induced by Space Flight

RAPD, one type of DNA molecular marker proposed by Willams et al (1990), was based on the electrophoretic band polymorphism generated by PCR amplification of genomic DNA from different sources using single primers 8 to 10 nucleotides in length. RAPD assay has become generally accepted as a powerful tool and widely used for plant genome analysis, such as gene mapping, genetic linkage map construction, germplasm identification, phylogenetic relationship determination etc. (Martin et al., 1991; Yu and Pauls, 1993; Howell et al., 1994). Report on the molecular analysis of DNA from long pod mutant of mung bean with RAPD has been published recently (Wang et al., 1996).

In the present study, we used one hundred and ten 10-mer ologonucleotides with arbitrary sequence as primers in the RAPD assay for screening the polymorphic products among ten plants of M_1 and M_2 generations from flown seeds with striped leaves and the control plant. Totally five hundred and thirteen products, i.e. visible bands on electrophoretic gel, were produced, and only 10.9% of primers generated 14 polymorphic bands between DNA from the mutant plants and those from control plants, as indicated in Table 2. If each band (product of PCR) was assumed to be one locus in the maize genome, the variation rate of loci in these mutated plants could be calculated to be 2.7%, according to the results from this study.

Table 2 DNA Polymorphism between Plants from Flown Seeds with Striped Leaves and Those from Ground Control

Primers screened	Primers generating polymorphic products		Total fragments produced	Polymorphic products	
No.	No.	(%)	No.	No.	(%)
110	12	10.9	513	14	2.7

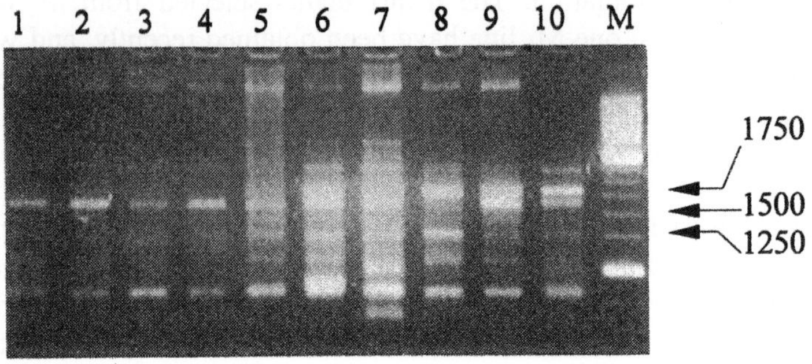

Fig. 1. PCR amplification directed by primer OPE-12 using DNA from the maize plants No. : * lane 1. 2-2; lane 2. 2-28; lane 3. 2-30; lane 4. 2-35; lane 5. 2-41; lane 6. 11; land 7. 11- 1; lane 8. 11-16; lane 9. 11-23; lane 10. Control; lane M. Molecular weight size standard, 250 bp ladder (from Phamacia Biotech)

(* Number in front of dash - No. of first generation plants from flown seeds with this number.
Number behind dash - No. of second generation plants in the selfed lines from the flown seeds)

Genomic DNAs from the selfed progenies developed from seeds 2 and seed 11, with large yellow stripes on leaves, were used as templates to explore the common polymorphic bands in one line with the same phenotypic mutation, when compared to control in the RAPD assay using the primers generated polymorphic bands. Of 12 primers surveyed, three primers, i.e. OPE-12, OPE-16, and OPS-13, were found to generate common variation in electrophoresis band patterns. For example, Figure 1 shows the amplification results using OPE-12 as primer to compare the band patterns of DNA from the progenies of seed 2, seed 11, and that of control plant. The band patterns appeared identical for all five progenies from seed 2 and among three of the four progenies from seed 11. The RAPD assay using DNA from plants with other mutated characters, i.e. dwarf or flower anormagensis, observed in the second generation, as templates and twenty 10-mers random primers was also performed. Preliminary results showed that variations in banding patterns correlated with specific mutation when a certain primer was used. The data suggested that RAPD assay might be used as an effective and economic tool in assessing genetic damage of plant systems induced by space flight.

It should be noted that the types and frequencies of mutation observed in the present study are similar to those from we reported previously (Mei et al., 1994), even though the length of these two flights was quite different. This similarity in biological response demonstrated that space radiation environment could induce inheritable mutagenic effects on plant seeds, and provided preliminary evidence in support of the report from China on space mutation breeding (Jiang, 1996). For a better understanding of the mechanism of mutation induced by space flight, further space flight experiments will be needed to study the radiation dosimetry in LEO, fine location of HZE particles in flown seeds, mutational events in several generations after flight, as well as cloning and sequencing the loci coding the specific phenotype examined.

ACKNOWLEDGEMENT

This study is supported by the National Natural Foundation of China Grant No. 39680007. We would like to thank Dr. Tracy C. Yang for his valuable suggestions to this study and manuscript, and Dr. Tom K. Hei for his critical review of this manuscript.

REFERENCES

Bucker, H. The Biostack Experiments I and I aboard Apollo16 and 17. *Life Sci. Space Res.*, **12**, 43-50 (1974)

Dellaporta, S.L. and Calderon-Urrea, A. The Sex Determination Process in Maize. *Science.*, **266**, 1501-1505 (1994)

Horneck, G. Radiobiological Experiments in Space: A Review. *Nucl. Tracks. Radiat. Meas.*, **20**, 185-205 (1992)

Howell, E.C., Newbury H.J., SwennenR.L., Withers, L.A. and Ford-Lord B.V. The Use of RAPD for Identifying and Classifying of *Musa* Germplasm. *Genome* **37**, 328-332 (1994)

Jiang, X. Development and Prospect of Space Mutation Breeding in China. *Chinese Journal of Space Science.,* **16** (Supp.), 77-82 (1996)

Martin, G.B., Williams, J.G.K. and Tanksley, S.D. Rapid Identification of Markers Linked to a *Pseudomonas* resistance gene in tomato by Using Random Primers and Near Isogenic Lines. *Proc. Natl. Acad. Sci.* USA **88**, 2336-2340 (1991)

Mei, M., Qiu, Y., He, Y., Bucker, H., and Yang, C.H. Mutational Effects of Space Flight on *Zea mays* Seeds. *Adv. Space Res.*, 14 (10), 33-39 (1994)

Murray, M. G. and Thompson, W.F Rapid Isolation of High Molecular Weight Plant DNA. *Nucleic Acid Res.* **8**, 4321-4325 (1980)

Peterson, D.D., Benton, E.V., Tran, M., Yang, T., Freeling, M., Craise, L., and Tobias, C. A. Biological Effects of High -LET Particles on Corn Seed Embryos in the Apollo-Soyuz Test Project Biostack III Experiment. *Life Sci. Space Res.*, **15**, 151-155 (1977)

Qiu, Y., Mei, M., He, Y. and Lu, Y. Mutagenic Effects of Accelerated Heavy Ion Irradiation on *Zea mays. J.South China Agricultural University.,* **12**, 48-54 (1991)

Wang, B., Li, J., Qiu, F., Wang, P., Han, D. and Jiang, X. Breeding by Space Mutagenesis in Mung Bean and it's Molecular Analysis. *Chinese Journal of Space Science*, **16**(Supp.), 121-124 (1996)

Williams, J. G. K., Kubelik, A.R., Livak, K.J., Rafalski, J.A. and Tingey, S.V. DNA Polymorphism Amplified by Arbitrary Primers are Useful as Genetic Markers. *Nucleic Acids Res.* **18**, 6531- 6535 (1990)

Yang, T. C. and Tobias, C. A. Potential Use of Heavy-Ion Radiation in Crop Improvement. *Gamma-Field Symposia,* No. 18, 141-151 (1979)

Yu, K. and Pauls, K.P. Identification of a RAPD Markers associated with Somatic Embryogenesis in Alfalfa. *Plant Mol. Biol.* **22**, 269-277 (1993)

Pergamon

Adv. Space Res. Vol. 22, No. 12, pp. 1699–1707, 1998

Printed in Great Britain
0273-1177/98 $19.00 + 0.00

PII: S0273-1177(99)00035-6

GENOMIC INSTABILITY AND TUMORIGENIC INDUCTION IN IMMORTALIZED HUMAN BRONCHIAL EPITHELIAL CELLS BY HEAVY IONS

T.K. Hei, C.Q. Piao, L.J. Wu, J.C. Willey[1], and E.J. Hall

Center for Radiological Research, College of Physicians & Surgeons, Columbia University, New York, NY., 10032 U.S.A.
[1]Department of Medicine, Medical College of Ohio, Toledo, OH., 43699 U.S.A.

ABSTRACT

Carcinogenesis is postulated to be a progressive multistage process characterized by an increase in genomic instability and clonal selection with each mutational event endowing a selective growth advantage. Genomic instability as manifested by the amplification of specific gene fragments is common among tumor and transformed cells. In the present study, immortalized human bronchial (BEP2D) cells were irradiated with graded doses of either 1GeV/nucleon ^{56}Fe ions or 150 keV/μm alpha particles. Transformed cells developed through a series of successive steps before becoming tumorigenic in nude mice. Tumorigenic cells showed neither *ras* mutations nor deletion in the p16 tumor suppressor gene. In contrast, they harbored mutations in the p53 gene and over-expressed cyclin D1. Genomic instability among transformed cells at various stage of the carcinogenic process was examined based on frequencies of PALA resistance. Incidence of genomic instability was highest among established tumor cell lines relative to transformed, non-tumorigenic and control cell lines. Treatment of BEP2D cells with a 4 mM dose of the aminothiol WR-1065 significantly reduced their neoplastic transforming response to ^{56}Fe particles. This model provides an opportunity to study the cellular and molecular mechanisms involved in malignant transformation of human epithelial cells by heavy ions.

Introduction

Carcinogenesis is a multi-stage process with sequences of genetic events governing the phenotypic expression of a series of transformation events leading to the development of metastatic cancer. An understanding of the carcinogenic mechanisms of high LET radiation is essential for human risk estimation and radiation protection. The carcinogenic risk for human epithelial cells after exposure to high LET radiation has been estimated to be in the range of 10^{-12}/cell/Gy based on the breast cancer incidence among Japanese A-bomb survivors (Hei *et al.*, 1996a). However, the mechanisms for radiation carcinogenesis by high LET radiation such as alpha particles and heavy ions are not clear. The use of lung tumor tissues from underground miners exposed to radon in identifying consistent cellular and molecular alterations, as in the case of human colorectal cancers, is complicated by the findings that the majority of miners are also cigarette smokers. It will be ideal to use a human bronchial epithelial cell line that has been malignantly transformed by radiation to assess the various changes leading to malignancies. *In vitro* oncogenic transformation studies using rodent cell systems have shown that high LET radiation is more efficient in transforming cells than X or γ-rays at equivalent doses with a relative biological effectiveness ranging from 2.2 to 10 (Hei *et al.*, 1988, Hall and Hei, 1985). Since human epithelial cells rarely undergo spontaneous immortalization and are extremely refractory to *in vitro* neoplastic transformation by carcinogens (Rhim, 1991 for review), model systems based on immortalized human epithelial cells are extremely valuable to provide qualitative data on mechanisms of radiation-induced carcinogenesis. This is particularly true when using ionizing radiation where large, multiple doses are often required either to immortalize cells (Namba *et al.*, 1986) or to convert previously immortalized cells to malignant cells (Yang *et al.*, 1991, Thraves *et al.*, 1990).

Immortalized Human Bronchial Epithelial Cell Model

To better understand the cellular and molecular mechanisms involved in human bronchial carcinogenesis induced by either heavy ions or environmental carcinogens such as asbestos fibers, we have developed a transformation model based on human papillomavirus immortalized human bronchial epithelial (BEP2D) cells as shown in Figure 1. BEP2D cells are initiated by lipofectin transfection of cloned full

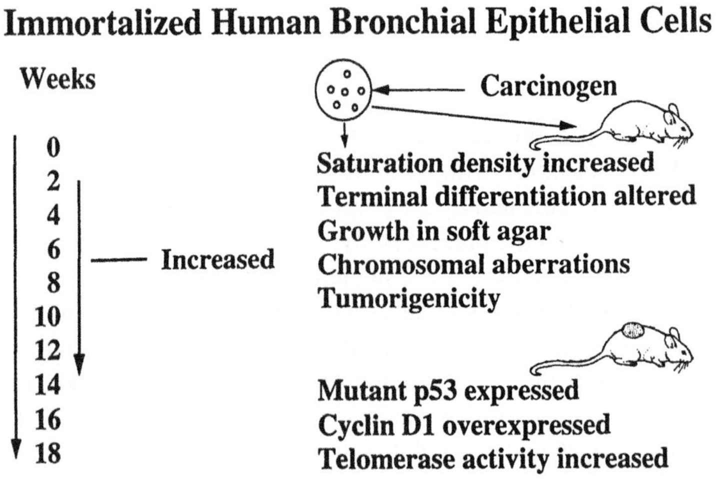

Figure 1. Schematic diagram illustrating the multistep process in the neoplastic transformation of immortalized human bronchial epithelial cells irradiated with a single 60 Gy dose of 1 GeV/nucleon ^{56}Fe ions. Irradiated cells need to undergo successive passages and the accumulation of additional phenotypic/ mutagenic changes before tumorigenicity can be demonstrated.

length HPV18 into normal human bronchial epithelial cells obtained as an outgrowth of bronchial explant (Willey *et al.*, 1991). Although these bronchial epithelial cells are immortal, they are anchorage dependent and do not form tumors in immunosuppressed host animals. After carcinogen treatment, transformed cells arise through a series of sequential stages including altered growth pattern, resistance to serum-induced terminal differentiation, agar-positive growth, tumorigenicity, and metastasis (Hei *et al.*, 1994a, 1996a & b, 1997). It should be noted that, while the majority of agar-positive BEP2D clones are non-tumorigenic, they all demonstrated the propensity to resist serum-induced terminal differentiation. In addition, each preceding stage represents a necessary, yet insufficient step towards the later, more malignant phase. Northern and Western blot analyses showed an over-expression of cyclin D1 (Hei *et al.*, 1996b) and mutated p53 oncoproteins (Hei *et al.*, 1996a) among the tumorigenic BEP2D cells. Since chromosome end to end associations and telomerase activity are often associated with genomic instability and cellular immortality respectively, our recent data which show an overabundant increase in telomerase activity together with the highest frequency of chromosome end associations among the metastatic cell lines, suggest that they may be useful indicators of metastatic potential in radiation induced lung cancers (Pandita *et al.*, 1996).

Exponentially growing BEP2D cells plated in T25 tissue culture flasks were irradiated with graded doses of 1 GeV/nucleon ^{56}Fe ions accelerated with the Alternating Gradient Synchrotron at the Brookhaven National Laboratory. After irradiation, cells were trypsinized, counted, and replated for both survival and the expression of transformed phenotypes as described previously (Hei *et al.*, 1994a, 1996a & b). Irradiated cells demonstrated a dose dependent cytotoxicity with a mean lethal dose of 0.7 Gy as shown in Figure 2. Compared to the 0.2 Gy dose value obtained with 150 keV/µm ^4He ions, these very high energy particles were less effective in killing BEP2D cells under comparable culture conditions. The RBE for cell lethality at the D_0 dose was 6.0 and 3.4 for α-particles and ^{56}Fe ions respectively. Cells irradiated with either a 0.3 or 1 Gy dose of ^{56}Fe ions developed anchorage independent clones at frequencies ranging from 0.04 to 0.4% (Table I) similar to those previously reported after alpha particle irradiation (Hei *et al.*, 1996a). Upon inoculation into nude mice, only cells irradiated with the higher dose groups produced tumors (2/4 and 3/4 animals for the 0.6 and 1 Gy doses respectively).

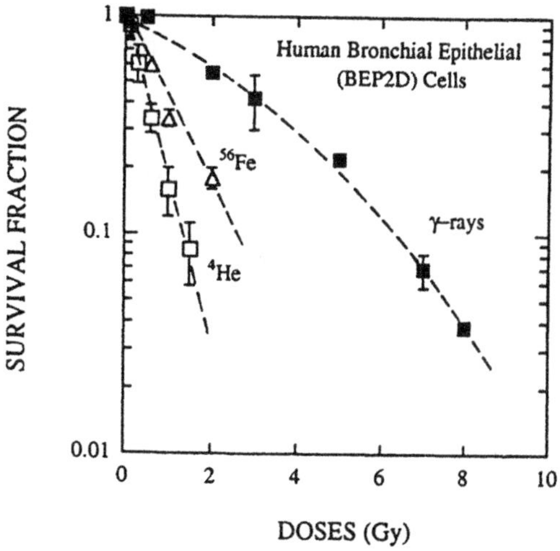

Figure 2. Survival fraction of BEP2D cells irradiated with graded doses of either 1 GeV/nucleon ^{56}Fe ions, 150 keV/µ alpha particles, or γ-rays. Data are pooled from 3-7 experiments. Bars represent ± SEM.

TABLE I

Neoplastic Transformation Incidence in BEP2D cells Irradiated with 1 GeV/nucleon ^{56}Fe Ions[1]

Dose (cGy)	Time in Culture (week)	Growth in Agar[2]	Tumorigenic incidence[3]
0	12	-	0/4[4]
30	12	+	0/4
60	12	+++	2/4
100	12	+++	3/4

[1] Total number of cells irradiated per group ranged from 2.3 to 12×10^6 cells

[2] Soft agar colonies ranged from 0.02 to 0.4%

[3] Animals with palpable nodules > 0.5 cm/ total number of animals injected

[4] Control BEP2D cells have not produced a single tumor among 47 animals injected including historical controls.

Evidence That Loss of Suppressor Gene Function Mediates Radiation-Induced Transformation of BEP2D cells

The development of recombinant DNA technology during the past decade has led to the recognition that cancer may be a result of either the activation of oncogene(s) or the loss of tumor suppressor genes. So far, no oncogene has been identified as the causal step in radiation induced tumor *in vivo*, or with radiation induced transformation *in vitro*. Although *ras* oncogenes were shown to be activated in certain X-ray induced lymphomas in mice (Guerrero *et al.*, 1984), this alteration was found in only a fraction of the tumors and might not represent the causal event. Likewise, consistent with our data with the BEP2D cells, no mutations in any of the *ras* oncogenes has been identified among radiation-induced tumorigenic human keratinocytes (Thraves *et al.*, 1990, 1995). The fact that heavy ions and other high LET radiation are efficient inducer of large chromosomal deletions (Evans *et al.*, 1991, 1998, Hei *et al.*, 1994b, Zhu *et al.*, 1996), provides a mechanism for the loss of suppressor functions. Studies with somatic cell hybrids have shown that tumor suppression occurs in neoplastic cells and can be corrected with cell fusion with normal human chromosomes (Sager, 1985, Saxon *et al.*, 1986). To understand the mechanism of radiation carcinogenesis, we carried out cell fusion studies to determine whether tumorigenicity of BEP2D cells behaves as a dominant or recessive trait as shown in Figure 3.

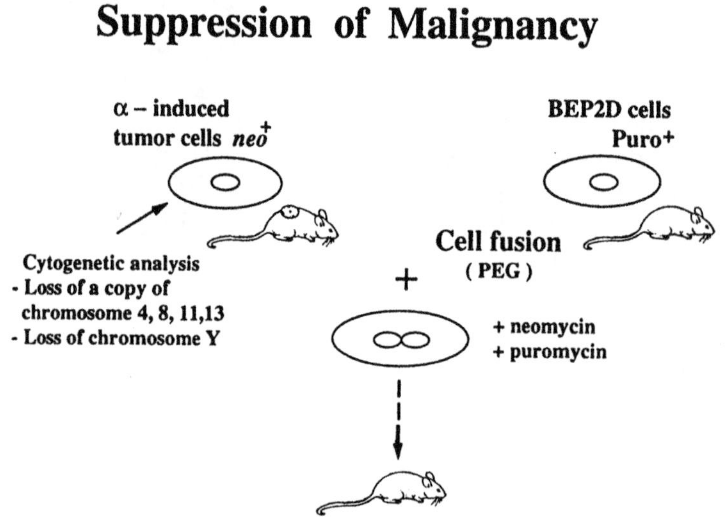

Figure 3. Cell fusion approach to examine the role of suppressor functions among tumorigenic BEP2D cells induced by alpha particles.

Exponential phase cultures of secondary BEP2D tumor cells induced by a single 60 cGy dose of alpha particles were transfected with the pRV/CMV expression vector containing a *neo* gene whereas control cells were stably transfected with the pBabe plasmid containing a puromycin resistant gene. 5 $\times 10^6$ tumorigenic cells were fused with an equal number of control BEP2D cells using polyethylene glycol 1500 applied in a drop-wise fashion. Resultant fusion cells were then selected in medium containing both G418 and puromycin over a 12 day period, expanded in culture, and re-inoculated into nude mice for tumorigenic expression. Results of these fusion experiments demonstrated that radiation-induced tumorigenic phenotype in BEP2D cells could be completely suppressed by fusion with non-tumorigenic

BEP2D cells. Furthermore, concurrent fusion of tumor cells to tumor cells resulted in tumorigenic hybrids whereas fusion among wild type BEP2D cells resulted in non-tumorigenic hybrid clones. These data indicate that non-tumorigenic BEP2D cells complement the loss of putative suppressor elements among tumorigenic cells and suggest that loss of suppressor gene(s) as a likely mechanism of radiation carcinogenesis.

Tumor Suppressor Gene and Genomic Instability

Studies on the functions of tumor suppressor genes have revealed that many of these gene products play a crucial role in the control of cellular growth and differentiation as well as in cell cycle control (Knudson, 1993, Muller *et al.*, 1993). Table II listed the many possible functions of tumor suppressor genes in mammalian cells including maintenance of genomic instability. A possibility that now seems more likely in cancer development is that tumor progression is a consequence of genomic instability and clonal selection, each mutation endowing a selective growth advantage. The notion is that a mutation may occur in a gene responsible for the stability of the genome and the fidelity of replication, resulting in what has been referred to as mutator phenotype, i.e. a single induced mutation followed by a cascade of further mutations. Support of this concept comes from the observation of microsatellite instability in a wide range of human tumors (Cheng & Loeb, 1993). The discovery of mutations in one of the five mismatch repair genes in cases of hereditary non-polyposis colorectal cancer, also support the idea that an induced mutation can result in instability and a mutator phenotype (Loeb, 1991). Although there is no clear cut evidence that genomic instability actually occurs among radiation induced cancer, the notion of global chromosomal instability arising from an initial event, possibly a mutation, makes it possible to understand conceptually how a single low dose of radiation can lead to a cancer many years after the initial exposure and seemingly involves multiple steps.

Table II

Possible Functions of Tumor Suppression Genes

Maintenance of genomic stability
Induce apoptosis
Induce differentiation and trigger cellular senescence
Regulate cell growth as a negative growth factor
Increase cellular communication
Inhibit proteolytic degradation of gene products involve in growth regulation

There is evidence that genomic instability contributes to the progression of tumorigenesis (Lengauer *et al.*, 1997). One aspect of genomic instability is gene amplification which is frequently observed in tumors and transformed cell lines. Amplification in the CAD gene which results in the acquired resistance to the chemotherapeutic agent PALA (N-phosphonacetyl-L-aspartate) has previously been demonstrated in rodent tumor cell lines (Tlsty *et al.*, 1989) but not in normal human fibroblasts (Wright *et al.*, 1990). Figure 4 shows the frequencies of PALA resistance among control, transformed but not yet tumorigenic, and tumorigenic BEP2D cells induced by a single 60 Gy dose of ^{56}Fe ions. CAD is a multifunctional protein that catalyzes the first three steps in the *de novo* biosynthesis of uridine monophosphate. PALA is a competitive inhibitor of the enzyme aspartate transcarbamylase (ATCase). In the present study, cultures were exposed to 9x LD_{50} concentration of PALA (180-200 μM) for 3 weeks to assess the frequency of drug-resistant clones. The frequency of PALA resistance among the immortalized, control BEP2D cells was less than 10^{-7}. In contrast, the frequency of gene amplification in the five tumor cell lines ranged from 1-3 $\times 10^{-3}$ and between 7-9 $\times 10^{-5}$ among the four transformed cell lines examined. The results demonstrate that the step-wise neoplastic transformation process induced

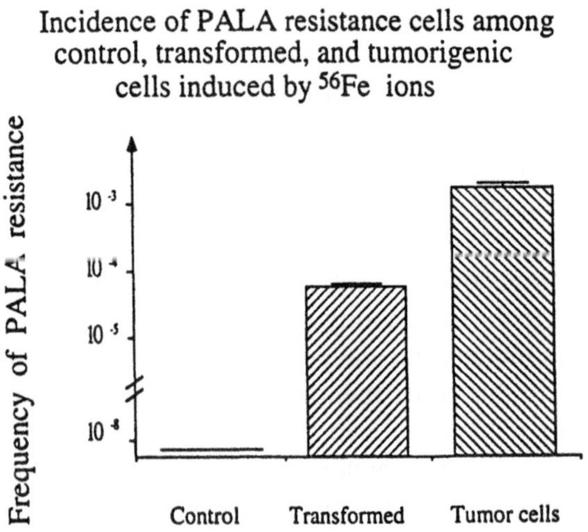

Figure 4. Frequency of PALA resistance among control, transformed, and tumorigenic BEP2D cells induced by heavy ions. Data are pooled from 3-5 experiments. Bars represent ± SEM.

by heavy ions is clearly associated with a gradual increase in genomic instability as determined by CAD gene amplification. Although the initial molecular events leading to gene amplification are not known, there is evidence to suggest that chromosomal breakage followed by formation of acentric fragments which harbor the target gene may play a role (Biedler *et al.*, 1988, Windle and Wahl, 1992). The significant increase in PALA resistance among the tumorigenic compared to control BEP2D cells may be useful as a predictive assay for tumorigenicity in BEP2D cells transformed by heavy ions.

Radiation Protection Studies with WR-1065 using BEP2D Cell Assay

Since the beginning of manned flight into space officially commenced in the sixties, the potential health hazard from exposure to the natural radiation environment outside the magnetic shielding of the earth has been a major concern of various space agencies including NASA. Although heavy ions constitute only a small percentage of the radiation field in outer space, they are thought to have a significant impact on the perceived risk. The ICRP has assigned a weighting factor of 20 for heavy ions, i.e. the risk of induced cancer from a given dose of heavy ions is 20x the risk from equivalent dose of low LET radiation (Sinclair, 1994). With the planned international space station program underway, information on realistic risk assessment and radiation protection is urgently needed.

2-(aminopropyl)-aminoethanethiol (WR1065) is the corresponding free thiol of the well-characterized radioprotector, Amifostine (WR-2721). Amifostine is a pro-drug which needs to be metabolized to its thiol form, WR-1065, before it is functionally active as a radioprotector (Grdina *et al.*, 1995). There is evidence from both *in vitro* and *in vivo* studies that WR-2721 and WR-1065 are anticarcinogenic, antimutagenic, and protect against radiation and cis-platinum induced mutagenesis in mammalian cells (Hill *et al.*, 1986, Milas *et al.*, 1984, Grdina *et al.*, 1992). These studies form the basis for the current interest in Amifostine as a chemopreventive agent to be used as an adjuvant in chemotherapy and radiotherapy. Figure 5 shows the effects of a 4 mM dose of WR-1065 given 2 hr before and 2 hr after radiation on the malignant transformation incidence of BEP2D cells irradiated with a single 60 cGy dose of 1 GeV/nucleon ^{56}Fe ions. Control and irradiated cells with or without drug treatment were subcultured for a period of 10-12 weeks before being inoculated into nude mice for assessment of their tumorigenic potential in nude mice. Although pre-treatment with WR-1065 had no protective effect on clonogenic survival of ^{56}Fe ion-irradiated BEP2D cells (data not shown), it obliterated the tumorigenic potential of these cells upon inoculation in nude mice.

Successful chemopreventive drugs can target specific phase of the cancer developmental process such as enzymatic activation of chemical carcinogens, induce programmed cell death to functionally-altered cells, restore normal cell differentiation and tumor suppressor gene function. As such, chemopreventive drugs can be classified into several categories: anti-mutagenic (Phase II enzyme inducer oltipraz, polyphenols); anti-proliferative (retinoid, difluoromethylornithine); anti-inflammatory (aspirin, sulindac), and antioxidants (thiols, Amifostin). The precise mechanism for the anti-tumorigenic effects of WR-1065 in heavy ion-irradiated BEP2D cells is not clear. There is evidence that an increase in intracellular glutathione content following WR-1065 treatment may be important in mediating its anti-carcinogenic, anti-mutagenic effect *in vitro* (Grdina *et al.*, 1995).

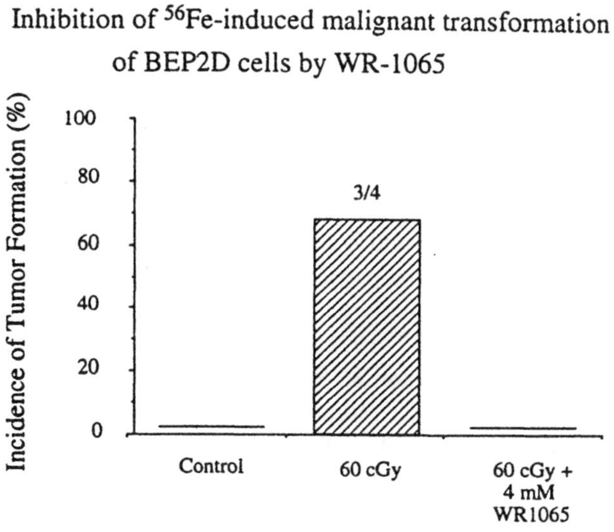

Figure 5. Tumorigenic incidence in nude mice injected with either control or ^{56}Fe ions irradiated BEP2D cells with or without pretreatment with a 4 mM dose of WR-1065 given 2 hr before and 2 hr after irradiation. 4-6 x10^6 cells in 0.2 ml phosphate buffered saline were injected per animal. Latency period averaged 8-10 weeks for irradiated BEP2D cells without pre-treatment with WR-1065. Animals injected with control and irradiated BEP2D cells pre-treated with drug were followed for period up to 8 months post-inoculation.

ACKNOWLEDGMENTS

Work supported by grants from the U.S. National Institute of Health CA 49062, CA/NASA 73946, ES 07890, and ES 05719.

REFERENCE

Biedler, J.L., T.D. Chang, K.W. Scotto, P.W. Melera, and B.A. Spendler, Chromosomal organization of amplified genes in multidrug-resistant Chinese Hamster Cells. *Cancer Res.*, **48**, 3179 (1988).

Cheng, K.C., and L.A. Loeb, Genomic instability and tumor progression. *Adv. Cancer Res.*, **60**, 121 (1993).

Evans, H.H., Cellular and molecular effects of radon and other alpha particles emitters, *Adv. Matagenesis Res.*, **3**, 29 (1991).

Evans, H.H., T.E. Evans, and M.F. Horng., Mutagenicity of ^{56}Fe irradiated at the TK locus in human lymphoblasts. Proc. Radiat. Res. Society Meeting, Louisville, Ky., 15-279 (1998)

Grdina, D.J., Y. Kataoka, I. Basic, and J. Perrin, The radioprotector WR2721 reduces neutron-induced mutations at the HPRT locus in mouse splenocytes when administered prior to or following irradiation. *Carcinogenesis* **13**, 811 (1992).

Grdina, D.J., N. Shigematsu, D. Phylis, G.L. Newton, J.A. Aguilera, and R.C. Fahey, Thiol and disulfide metabolites of the radiation protector and potential chemopreventive agent WR-2721 are linked to both its anti-cytotoxic and anti-mutagenic mechanisms of action, *Carcinogenesis* **16**, 767 (1995).

Guerrero, J., A., Villasonte, V. Corces, and A. Pellicer, Activation of a c-K-ras oncogene by somatic mutation in mouse lymphomas induced by gamma-radiation. *Science*, **225,** 1159 (1984).

Hall, E.J., and T.K. Hei, Oncogenic transformation with radiation and chemicals - A review. *Int. J. Radiat. Biol.* **48**, 1 (1985).

Hei, T.K., K. Komatsu, and E.J. Hall, Oncogenic transformation by charged particles of defined LET. *Carcinogenesis* **9**, 747 (1988)

Hei, T.K., C.Q.,Piao, J.C. Willey, S.Thomas, and E.J. Hall, Malignant transformation of human bronchial epithelial cells by radon-simulated alpha particles. *Carcinogenesis* **15**, 431 (1994a).

Hei, T.K., L.X. Zhu, D.Vannais, and C.A.Waldren, Molecular analysis of mutagenesis by high LET radiation. *Adv. Space Res.* 14 (10): 355 (1994b).

Hei, T.K., C.Q. Piao, T. Sutter, J.C. Willey, and K. Suzuki, Cellular and molecular alterations in human epithelial cells transformed by high LET radiation. *Adv. Space Res.* **18**, 137 (1996a).

Hei, T.K., C.Q. Piao, E. Han, T. Sutter, and J.C. Willey, Radon induced neoplastic transformation of human bronchial epithelial cells. *Radiation Oncology Investigation,* 3, 398 (1996b).

Hei, T.K., L.J. Wu, and C.Q. Piao, Malignant transformation of immortalized human bronchial epithelial cells by asbestos fibers. *Environmental Hlth. Persp.***105**, 1085 (1997).

Hill, C.K., B., Nagy, C. Peraino, and D.J. Grdina, WR-1065 is anti-neoplastic and anti-mutagenic when given during ^{60}Co-γ rays irradiation. *Carcinogenesis* **7**, 665 (1986).

Knudsen, A.G., Jr., Antioncogenes and human cancer: a review. *Proc. Natl. Acad. Sci.*, **90**,10914 (1993).

Lengauer, C., K.W. Kinzler, and B. Vogelstein, Genetic instability in colorectal cancers, *Nature*, **386**, 623 (1997).

Loeb, L.A., Mutator phenotype may be required for multistage carcinogenesis. *Cancer Res.*, **51**: 3075 (1991).

Milas, L., N. Hunter, C.L. Stephens, and L.J. Peters, Inhibition of radiation carcinogenesis by WR-2721. *Cancer Res.,* **44,** 5567 (1984).

Muller, C.A., R.Mumberg, and F.C. Lucibello, Signals and genes in the control of cell cycle progression. *Biochem. et Biophys. Acta,* **1155,** 151 (1993).

Namba, M., K. Nishitani, F. Fukushima, T. Kimoto, and K. Nose, Multistep process of neoplastic transformation of normal human fibroblasts by ^{60}Co gamma-rays and Harvey sarcoma viruses, *Int. J. Cancer,* **37,** 419 (1986).

Pandita, T.J., E.J. Hall, T.K. Hei, M.A. Piatyszek, W.E. Wright, C.Q. Piao, R.K. Pandita, J.C. Willey, C.R. Geard, and J.W. Shay, Chromosome end to end associations and telomerase activity during cancer progression in human cells after treatment with alpha particles simulating radon progeny. *Oncogene* **13,** 1423 (1996).

Rhim, J.S. and A. Dritschilo, Neoplastic transformation in human cell system-an overview, In: *Neoplastic Transformation in Human Cell Culture: Mechanisms of Carcinogenesis,* J.S.Rhim and A. Dritschilo eds., pp. Xi-xxxi, Totowa, NJ., Humana Press (1991).

Saxon, P.J., E.S. Srivatsan, and E.J. Stanbridge, Introduction of chromsome 11 via microcell transfer controls tumorigenic expression of HeLa cells. *EMBO,* **4,** 3461 (1986).

Sager, R., Genetic suppression of tumor formation. *Adv. Cancer Res.,* **44,** 43 (1985).

Sinclair, W.K., Radiation protection issues in galactic cosmic ray risk assessment. *Adv. Space Res.* **14**(10), 879, 1994.

Thraves, P., Z. Salehi, A. Dritschilo, and J.S. Rhim, Neoplastic transformation of immortalized human epidermal keratinocytes by ionizing radiation, *Proc. Natl. Acad. Sci.,* **87,** 1174 (1990).

Thraves, P., S. Varghese, M. Jung, D.J. Grdina, J.S. Rhim, and A. Dritschilo, Transformation of human epidermal keratinocytes with fission neutrons. *Carcinogenesis* **15,** 2867 (1995).

Tlsty, T.D., B. Margolin, K. Lum, Difference in the rates of gene amplification in non-tumorigenic and tumorigenic cell lines as measured by Luria-Delbruck fluctuation analysis. *Proc. Natl. Acad. Sci.,* **86,** 9441 (1989).

Willey, J.C., A. Broussoud, A. Sleemi, W.P. Bennett, P. Cerutti, and C.C. Curtis, Immortalization of human bronchial epithelial cells by papillomaviruses 16 or 18, *Cancer Res.,* **51,** 5370 (1991).

Windle, B.E., and G.M. Wahl, Molecular dissection of mammalian gene amplification: new mechanistic insights revealed by analyses of very early events. *Mutation Res.,* **276**(3), 199 (1992)

Wright, J.A., H.S. Smith, F.M. Watt, M.C. Hancock, D.L. Hudson, and G.R. Stark., DNA amplification is rare in normal human cells. *Proc. Natl. Acad. Sci.,* **87,** 1791 (1990).

Yang, T.C., M.R. Stampfer, and J.S. Rhim, Neoplastic transformation of human epithelial cells by ionizing radiation, In: *Neoplastic Transformation in Human Cell Culture: Mechanisms of Carcinogenesis,* J.S. Rhim and A. Dritschilo eds., (pp 103-111), Totowa, NJ: Humana Press (1991).

Zhu, L.X., C.A. Waldren, D. Vannais, and T.K. Hei, Cellular and molecular analysis of mutagenesis induced by charged particles of defined LET. *Radiation Res.,* **145,** 251 (1996).

Pergamon

Adv. Space Res. Vol. 22, No. 12, pp. 1709–1717, 1998

Printed in Great Britain
0273-1177/98 $19.00 + 0.00

PII: S0273–1177(99)00036–8

MICROSATELLITE INSTABILITY IN HUMAN MAMMARY EPITHELIAL CELLS TRANSFORMED BY HEAVY IONS

S.Yamada[1,2], T.C.Yang[2], K.George[2], R.Okayasu[3], K. Ando[1], and H.Tsujii[1]

[1]*Research Center of Charged Particle Therapy, National Institute of Radiological Sciences, Chiba263-8555, Japan,* [2] *NASA Johnson Space Center, Houston,TX77058, USA,* [3] *University of Texas Medical Branch, Galveston,TX77555, USA*

ABSTRACT

We analyzed DNA and proteins obtained from normal and transformed human mammary epithelial cells for studying the neoplastic transformation by high-LET irradiation *in vitro*. We also examined microsatellite instability in human mammary cells transformed to various stages of carcinogenesis, such as normal, growth variant and tumorigenic, using microsatellite marker D5S177 on the chromosome 5 and CY17 on the Chromosome 10. Microsatellite instabilities were detected in the tumorigenic stage. These results suggest that microsatellite instability may play a role in the progression of tumorigenecity. The cause of the genomic instability has been suggested as abnormalities of DNA-repair systems which may be due to one of the three reasons: 1) alterations of cell cycle regulating genes. 2) mutations in any of the DNA mismatch repair genes, 3) mutation in any of the DNA strand breaks repair genes. No abnormality of these genes and encoded proteins, however was found in the present studies. These studies thus suggest that the microsatellite instability is induced by an alternative mechanism.

INTRODUCTION

Recently, Kadhim and her coworkers(Kadhim et al. 1995) have reported a high frequency of non-clonal cytogenetic abnormalities in the clonal descendants of murine and human haemopoietic stem cell surviving low doses of high-LET irradiation. The data are compatible with the types of lesions induced by high-LET irradiation in stem cells that result in the transmission of genomic instability to their progeny, a characteristic often attributed to high-LET irradiation. Yang and Craise(1994) demonstrated that high-LET particles efficiently induce cell lethality and neoplastic transformation in both rodent and human cell assay systems. It is of great interest to establish relationship between genomicinstability and high-LET radiation induced DNA damagein human cells.

We further examine abnormalities of repair genes and cell cycle regulating genes in these transformants. For radiation risk assessment of long term flights to the Moon and Mars, a basic understanding of the mechanism(s) of carcinogenesis of charged particles is essential, particularly in terms of the relationship between genomic instability and high-LET radiation induced DNA damages to genes critical for human cellular functions.

METHODOLOGY

Cell System

Human mammary epithelial cells(**H184B5**) used originates from the laboratory of Martha Stampfer. They are immortal and non-tumorigenic and require special medium enriched with growth factors to grow. B5 cells were irradiated by a 2.2 Gy dose of iron particles(600MeV/μ) accelerated in Lawrence Berkeley Laboratory and selected for growth variants (**H185B5-F5/M4, M10**). In normal MEM medium supplemented with 10% fetal bovine serum, growth variants proliferated steadily while B5 cells senesce. We have obtained anchorage independent transformants (**H185B5F5/C#3, OC1-1, OC1-2**), which can grow in soft agar media, by irradiating the growth variants again with a second 2.2 Gy dose of iron beam.

We analyzed DNA obtained from both normal and transformed human mammary epithelial cell induced by high-LET radiation *in vitro* and compared with the data obtained **AG**1522 normal human fibroblast.

Microsatellite Analysis

The DNA samples were examined for genetic alterations at 22 different microsatellite loci by PCR amplifications as described by Dams et. al. (Dams et. al. 1995). PCR conditions consisted of a denaturation step for 10 min. at 94°C and 30-35 cycles at 94°C for 40s, 54-62°C for 30s and 72°C for 60s. Amplified PCR products were electrophoresed on a 6% polyacrylamide gel containing 6 M urea. The gel was run at room temperature for 4h at 1200V.

PCR-SSCP Analysis

PCR-SSCP method was performed for analysis of p53 gene as described by Balcer-Kubiczek et. al (Balcer-Kubiczek et.al.1995). Amplified PCR products were electrophoresed on a 6% polyacrylamide gel containing 5% glycerol. The gel was run for 12-16h at 400V.

Western Blot Analysis

Western Blots were performed to analyze p53, Rb, MSH2, MLH1, PMS2, DNA-PK, Ku70 and Ku86. Electrophoresis was performed on 5-30 mg protein per sample on 6-14% SDS-polyacylamide gels. The separated gels were transferred to nitrocellulose membranes for 24 hours at 4°C. The primary antibodies used were p53/DO-1, Rb/IF8, PMS2/C-20 (SantaCruz Biotechnology, CA),

MSH2/Ab-1, MLH1/Ab-1(Oncogene Science Inc, MA), Ku Ab-2/clone 111 for Ku 80, Ku Ab-4/cloneN3H10 for Ku 70 (Lab Vision CA) and DNA-PK/C19 (SantaCruz Biotechnology, CA). Subsequently the membranes were washed and incubated in ECL Western blotting detection reagent (Amersham RPN 2106) for 1 minute.

Table 1 Summary of Microsatellite Instability Data of Human Mammary Epithelial Cells

Locus	**H184B5M4** (growth Variant)	**H184B5F5C#3** (Transformant)
001/002(1q21)	○	○
D2S123(2p16)	○	○
D3S1300(3p14.2)	○	○
D3S1234(3p14.2)	○	○
EABMD(3p24)	○	○
D3S1244(3p24)	○	○
D5S107(5q11)	○	○
D5S117(5p15)	○	●
CY11(10q23)	○	●
D10S538(10p11)	○	○
D11S899(11S15.1) →	○	○
D11S955E(11S15.1)	○	○
D11S4677(11S15.1)	○	○
D13S218(13p13)	○	○
D13S160(13p13)	○	○
D13S267(13q12)	○	○
TP53.5(17q13.1)	○	○
TP53.6(17p13.2)	○	○
D17S261(17p11.2)	○	○
D17S1286(17p11.2)	○	○
D17S579(17p11.2)	○	○
THRA(17p11.2)	○	○

(○ no instability, ● instability)

RESULTS

We first investigated microsatellite instabilities by PCR analysis. Table 1 shows summary of the data obtained from human mammary epithelial cells. PCR analysis was repeated at least twice. We analyzed 21 microsatellite loci. Some of these loci have previously been shown to have instability in a high percentage of breast cancer, and some are located in the lesion of tumor suppressor genes such as BRCA(Futreal et al. 1994), FHIT(Ohta et al. 1996), PTEN(Li et al. 1997) and TSG101(Zhong et al. 1997). Mutations in microsatellite loci were identified by a change in size of the PCR-amplified allelic markers caused by an alteration in the number of repeats of basic microsatellite sequence. Of the 21 microsatellite loci analyzed, alterations were observed in two loci.

Figure 1A shows microsatellite instability in human mammary epithelial cells at D5S117 on chromosome 5. D5S117 may be located in dicentric chromosome involving 5 and 19 by G-banding in the M10 growth variant when compared with control cells (Durante et al. 1996). Transformants such as C#3 and OC1-1 were clearly altered for the dinucleotide repeat at D5S117.

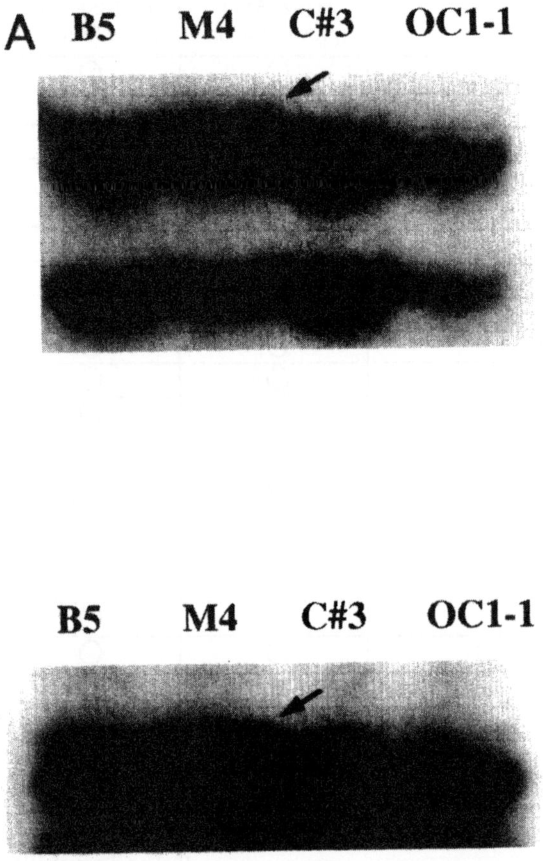

Fig.1 Microsatellite instability in Human Mammary Epithelial cells. DNAs were amplified with D5S117 on chromosome5(A) and with CY17 on Chromosome 10(B). Arrows, the position of alleles with electrophoretic mobility differences from the C#3 and OCI-1 when compared to the corresponding DNA of the B5 and M10. Upper band shifted to under band. Under band shows a strong signal. C#3 and OC1-1 demonstrate similar instability at both loci.

Figure 1B shows microsatellite instability in human mammary epithelial cells at CY17 on chromosome 10. A probe and is mapped to 10q23, near PTEN(Li et al. 1997). PTEN is a tumor suppressor gene located on chromosome band 10q23 and was isolated in 1997. It is often found to be mutated in sporadic brain, breast, and prostate cancer. Both of two transformed clone showed same instabilities in two loci. Analysis with whole chromosome painting probes revealed no alteration patterns between transformed and nontransformed clone (Yang et al. 1996). Moreover the location of the genomic instability is not random but specific. This nonrandom genomic instability can be considered as a late consequence of high-LET radiation (Martins et al. 1993). These results suggest that microsatellite instability may play a role in the progression of transformation.

It has been suggested that genomic instability may be caused by abnormalities of DNA-repair systems which may be due to one of the following three reasons: 1)alterations of cell cycle regulating genes, 2)mutations in any of the DNA mismatch repair genes, 3)mutations in any of the DNA strand breaks repair genes.

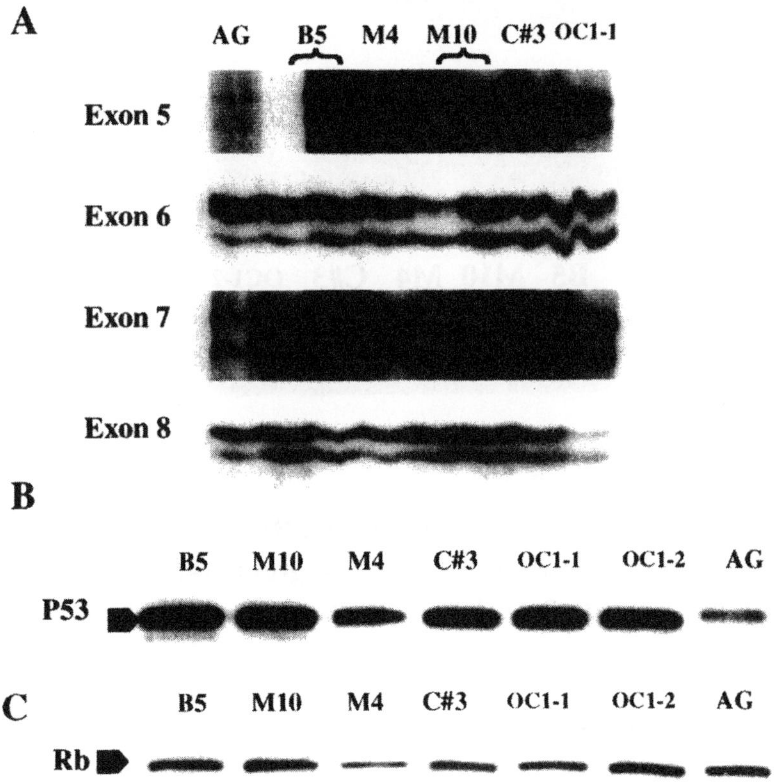

Fig.2 p53 and Rb status of Human Mammary Epithelial cells (A)PCR-SSCP analysis of the p53 gene, (B)Western blot analysis of p53 protein, (C) Western blot analysis of Rb protein. There is no mutation in p53 gene and no change in the expression of p53 and Rb protein.

1)` Alterations of cell cycle regulating genes

Abnormalities in the p53 or Rb have been shown to affect cell cycle control and lead to genomic instability in cell lines of either murine and human origin (Livingstone et al. 1992, Eyfjord et al. 1995). Normal cells arrest in G1 before entering S phase in response to DNA damage. This appears to be an important checkpoint to allow necessary DNA repair to take place. Cells lacking normal p53 and Rb do not show G1 arrest in response to DNA damage, and may therefore lead to an accumulation of unrepaired lesions and an increased mutation frequency.

We did not find any mutation in transformant clones when compared to normal and growth variants at the genomic DNA level. Furthermore, there was no alteration in either the p53 and Rb protein level. (Fig.2).

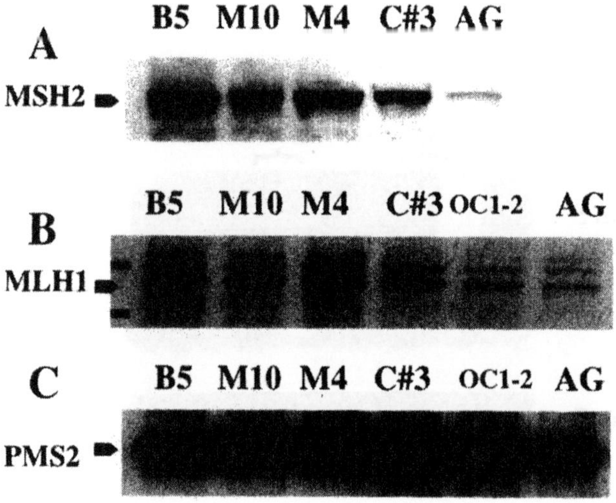

Fig.3 Western blot analysis of MSH2(A), MLH1(B) and PMS2(C) in normal human mammary epithelial cells(B5), growth variant(M10,M4), and transformants(C#3, OC1-1) and normal human fibroblast(AG1522).

2) Mutations in the DNA mismatch repair genes

DNA repair systems are essential in maintaining the structural integrity of genes. Unrepaired DNA damage may result in far-reaching consequences such as mutagenesis, genomic instability, tumorigenesis, and cell death (Orth et al. 1994). Genomic studies have demonstrated that the major DNA mismatch repair pathway requires MSH2, MLH1, and PMS2 which are likely to form a ternary complex during the initiation of eukaryotic DNA mismatch repair. Mutations in these gene can cause genomicinstability as manifested by an increased incidenceof microsatellite stability (Karran 1996).

Previous studies have conclusively shown that microsatellite instability in tumors arising in patients with HNPCC (hereditary nonpolyposis colorectal carcinoma) is due to the presence of mutations in one of three known mismatch repair genes (Liu et al. 1995). Our analysis revealed no change in the expression of MSH2, MLH1, and PMS2 in the transformed human mammary epithelial cells (Fig.3). These results suggest that instability of simple repetitive DNA sequences in the human mammary epithelial cells is not caused by mutations in the three mismatch repair gene above.

3) **Mutations in the DNA strand breaks repair genes**

Other forms of genomic instability are mediated by recombination . Recombination can be caused by errors in the repair of DNA double-strand breaks. Gross chromosomal rearrangements such as fusions, translocations, inversions and deletions are by definition recombination events. These events represent the breakage of chromosomal DNA followed by new rejoining in new and aberrant combinations (Phillips and Morgan 1994). As such, they can be thought of as errors in the repair of DNA double-strand breaks. DNA-PK (DNA-dependent protein kinase) comprises a catalytic subunit of a 460 kDa protein called DNA-PKcs and a DNA-binding component termed Ku, a heterodimer of 70 and 86 kDa subunits. In our studies the Ku70 and Ku86 were present in normal amounts (Fig.4). We have found no change in expression of DNA-PKcs protein.

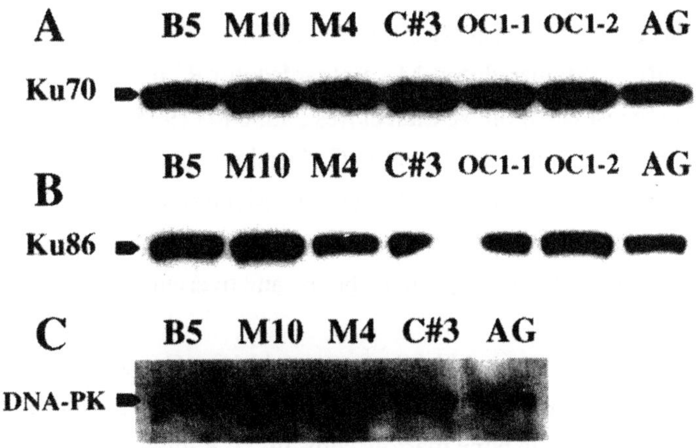

Fig.4 Western blot analysis of Ku70(A), Ku86(B) and DNA-PK(C) in normal human mammary epithelial cells(B5), growth variant(M10,M4), and transformants(C#3, OC1-1) and normal human fibroblast(AG1522).

SUMMARY

1. Microsatellite instability was detected in transformed human breast epithelial cells induced by high LET radiation using microsatellite marker D5S117 on the chromosome 5 and CY17 on the Chromosome 10.
2. No abnormality was found in studies with
   1) cell cycle regulating proteins p53 and Rb
   2) DNA mismatch repair proteins MSH2, MLH1 and PMS2
   3) DNA strand breaks repair proteins Ku 70, Ku86 and DNA-PK

These studies suggest that high-LET heavy ions induces microsatellite instability and that further studies of the mechanism(s) are needed.

References

Balcer-Kubiczek,M., Tin,J., Lin,K., Harrison,G.H., Abreham,J.M. and Meltzer, S.J.p53 mutation status and survival of human breast cancer MCF-7 cell variants after exposure to X rays or fission neutrons, Radiat. Res. 142,256(1995)

Dams, E., Van de Kelft, E.J.Z., Martin, J-J., Verlooy, J., and Willems, P.J. Instability of microsatellites in human gliomas, Cancer Res., 55,1547(1995)

Durante, M., Grossi, G. F. and Yang T. C., Radiation-induced chromosomal instability in human mammary epithelial cells, Adv. Space Res., 18, 99 (1996)

Eyfjord, J.E., Thorlacius, S., Steinarsdottir, M., Valgardsdottir,R., p53 Abnormalities and genetic instability in human breast carcinomas, Cancer Res., 55, 646 (1995).

Futreal,PA., Liu,Q., Shattuck-Eidens,D., Cochran,C., Harshman,K., Tavtigian,S., Bennett,LM., Haugen-Strano,A., Swensen,J., Miki,Y., Eddington,K.,Mcclure,M., Frye,C., Weaver-Feldhaus,J., Ding,W., Gholami,Z., Soderkvist,P., Terry,L., Jhanwar,S., Barrett,JC., Scolinick,HM., Kamb,A., and Wiseman,R., BRCA1 mutations in primary breast and ovarian carcinomas. Science, 266,120(1994)

Kadhim,M.A., Lorimore,S.A., Townsend,M.S., Townsnd,M.S., Goodhead,D.T., Bickle,V.J., and Wright,E.G., Radiation-induced genomic instability: delayed cytogenetic aberration and apoptosis in primary human bone marrow cells, Int. J. Radiat. Biol., 67,287 (1995).

Karran,P., Microsatellite instability and DNA mismatch repair in human cancer, Seminar in Cancer Biology 7, 15 (1996).

Li, J., Yen, C., Liaw, D., Podsypanina, K., Bose, S., Wang, S-I., Puc, J., Miliaresis, C., Rodgers, L., McCombie, R., Bigner, S. H., Giovenella, B. C., Ittmann, M., Tycko, B., Hibshoosh, H., Wigler, M.H. and Parsons, R., PTEN, a putative protein tyrosine phosphatase gene mutated in human brain, breast and prostate cancer, Science275, 1943 (1997)

Liu,B., Nicolaides,N.C., Markowitz,S., Willson,J.K.V., Parsons,R.E., Jen,J. Peltomaki,P., de la Chappelle,A., Hamilton,S.R., Kinzler,K.W., and Vogelstein,B., Mismatch repairgene defects in sporadic colorectal cancers with microsatellite instability, Nat. Genet., 9, 48 (1995).

Livingstone,L.R., White,A., Sprouse,J., Livanos,E., Jacks,T., and Tlsty,T.D., Altered cell cycle arrest and gene amplification potential accompany loss od wild-type P53, Cell,70,923 (1992).

Martins, M. B., Sabatier, L. and Ricoul, M., Specific chromosome instability induced by heavy ions, Mutation Res., 285, 229 (1993)

Ohta, M., Inoue H., Cotticelli, M.G., Kastury, K., Baffa, R., Palazzo, J., Siprashvili, Z., Mori, M., McCue, P., Druck, T., Croce C. M. and Huebner, K. The FHIT gene, spnning the chromosome 3p14.2 fragilesite and renal carcinoma-associated t(3:8) breakpoint, is abnormal in digestive tract cancers, Cell,84,587(1996)

Orth,K., Hung J., Gazdar A., Boecock, A., Genetic instability in human ovarian cancer cell lines, Proc. Natl. Acad. Sci. USA, 91, 9495, (1994).

Phillips,J.W. and W.F. Morgan, Illegitimaterecombination induced by DNA double-strand breaks in a mammalianchromosome, Molecular and cellularBiology, 14, 5794 (1994).

Yang,T. C-H.and Craise, L. M., Development of human epithelial cell systems for radiation risk assessment, Adv. Space Res., 14, 11 (1994).

Yang,T. C-H. George, K., Tavakoli, A. ,Craise, L.M. and Durante M., Radiogenictransformation of human mammary epithelial cells in vitro, Radiation Oncology Investigations, 3, 412 (1996)

Zhong Q. Chen CF. Chen Y. Chen PL. Lee WH. Identification of cellularTSG101 protein in multiple human breast cancer cell lines. [Journal Article] Cancer Res.,57,4225(1997).

Pergamon

Adv. Space Res. Vol. 22, No. 12, pp. 1719–1723, 1998

Printed in Great Britain
0273-1177/98 $19.00 + 0.00

PII: S0273-1177(99)00037-X

EFFECT OF TRACK STRUCTURE AND RADIOPROTECTORS ON THE INDUCTION OF ONCOGENIC TRANSFORMATION IN MURINE FIBROBLASTS BY HEAVY IONS

R. C. Miller[1], S. G. Martin[2], W. R. Hanson[3], S. A. Marino[1] and E. J. Hall[1]

[1]*Center for Radiological Research, Columbia University, New York, NY 10032*
[2]*CRC Dept. of Clinical Oncology, The University of Nottingham, City Hospital, Nottingham, U. K.*
[3]*Loyola-Hines Dept. of Radiotherapy, Loyola University, Hines, IL 60141*

ABSTRACT

The oncogenic potential of high-energy ^{56}Fe particles (1 GeV/nucleon) accelerated with the Alternating Gradient Synchrotron at the Brookhaven National Laboratory was examined utilizing the mouse C3H 10T1/2 cell model. The dose-averaged LET for high-energy ^{56}Fe is estimated to be 143 keV/μm with the exposure conditions used in this study. For ^{56}Fe ions, the maximum relative biological effectiveness (RBE_{max}) values for cell survival and oncogenic transformation were 7.71 and 16.5 respectively. Compared to 150 keV/μm ^{4}He nuclei, high–energy ^{56}Fe nuclei were significantly less effective in cell killing and oncogenic induction. The prostaglandin E_1 analog misoprostol, an effective oncoprotector of C3H 10T1/2 cells exposed to X rays, was evaluated for its potential as a radioprotector of oncogenic transformation with high-energy ^{56}Fe. Exposure of cells to misoprostol did not alter ^{56}Fe cytotoxicity or the rate of ^{56}Fe-induced oncogenic transformation.

INTRODUCTION

The environment outside earth's magnetic shield is a complex mixture of high-energy protons, electrons, alpha particles and heavy ions. The short- and long-term health effects to astronauts exposed to ionizing radiation during extended space exploration are of paramount concern to NASA. The potential for significant exposure to high-LET particles from manned space missions raises concern for both stochastic effects (cancer induction and heritable disorders) and deterministic effects (primarily cataract induction).

Cytotoxicity, chromosome aberrations and oncogenic transformation induced by heavy ions have provided useful data for estimating the health risks from a variety of radiations (Yang *et al.*, 1985; Miller *et al.*, 1995; Wu *et al.*, 1997). Based on the risk estimates, consideration must be given to agents that could ameliorate the radiation health hazards inherent in manned space flights. Misoprostol, a prostaglandin E_1 analog (Cytotec, G. D. Searle) has been shown to be an effective cytoprotector and oncoprotector of cells exposed to X rays and fission-spectrum neutrons (Hanson *et al.*, 1988; Hanson and Grdina, 1991; LaNasa *et al.*, 1994).

In the present study, the potential of misoprostol to protect astronauts from inevitable space radiation was examined using C3H 10T1/2 cells exposed to high-velocity ^{56}Fe ions.

MATERIALS & METHODS

Cell Culture

The C3H 10T1/2 cell line has been used successfully for many years to quantify the cytotoxic and oncogenic potential of a variety of radiations with LET values ranging from 0.25 to over 600 keV/μm (Yang *et al.*, 1985; Miller *et al.*, 1995). This model system is quantitative and has generated results that are highly reproducible. Cells from passages 8-14 were used throughout the study and were cultured and assessed for transformed colonies as described (Miller *et al.*, 1995). Two days before exposure, cells were plated into culture flasks. Because of the nature of beam availability from the synchrotron at the Brookhaven National Laboratory (BNL), cells treated with misoprostol were exposed to the drug beginning 12 hr before irradiation and removed 2-3 hr post-irradiation at which time cultures were processed for survival and oncogenic transformation.

Misoprostol

Prostaglandins and leukotrienes are end products of the arachidonic acid metabolism cascade. Among their broad array of effects, prostaglandins have been shown to protect tissues from a variety of injurious agents (Hanson *et al.*, 1988; Hanson and Grdina, 1991). In addition, prostaglandins have been shown in several cell systems to be potent radioprotectors of cells exposed to X rays, gamma rays and fission-spectrum neutrons. In order to be maximally effective, cells must be treated with misoprostol for at least 2 hr before exposure to radiation. In preliminary studies, C3H 10T1/2 cells displayed a significant increase in survival when a 5μg/ml dose of misoprostol was administered for at least 2 hr before exposure to X rays.

Irradiation Sources

The Alternating Gradient Synchrotron (AGS) at BNL provided nominal 1 GeV/nucleon ^{56}Fe ions. Based on a comprehensive examination of the beam, Zeitlin *et al.* (1998) reported the dose-averaged LET with the irradiation setup that we used was 143 keV/μm. Exposure conditions including prolonged exposures to misoprostol were employed for cells exposed to 250 kVp X rays or high-energy ^{56}Fe. The source of X rays was a Westinghouse Coronado X-ray unit operated at 250 kVp, 15 mA with 0.25mm Cu and 1 mm Al external filtration that provided an output of 1.14 Gy/min. The Van de Graaff accelerator at the Radiological Research Accelerator Facility (RARAF) provided ^{4}He ions with an initial energy of 5.6 MeV (2.4 MeV in the cell) with a measured LET of 150 keV/μm. For the ^{4}He ions studies, cells were trypsinized from stock plastic culture dishes and plated at 2×10^5 cells/dish onto thin Mylar-bottomed (6 μm thickness) stainless-steel dishes two days before irradiation. The Mylar dishes containing exponentially growing cells in monolayer were placed in a radiation wheel capable of holding up to 20 dishes. Immediately after irradiation, cells were removed from the Mylar surface and processed for examination of cell survival and oncogenic transformation as described in the previous section on cell culture techniques.

RESULTS

Dose response survival curves to 250 kVp X rays, 150 keV/μm ^{4}He nuclei and 1 GeV/nucleon ^{56}Fe are shown in Figure 1. Cells exposed to X rays display an initial shallow slope (α) followed by a continuously bending curve at higher doses (β). In contrast, cells exposed to either 150 keV/μm ^{4}He ions or 1 GeV/nucleon ^{56}Fe exhibit a dose response relationship where survival is a simple exponential function of dose. If the initial slopes are compared ($\alpha_{He\ nuclei}/\alpha_{X\ rays}$ and $\alpha_{Fe\text{-}56}/\alpha_{X\ rays}$), the RBE_{max} values for survival are 15.1 and 7.71 respectively.

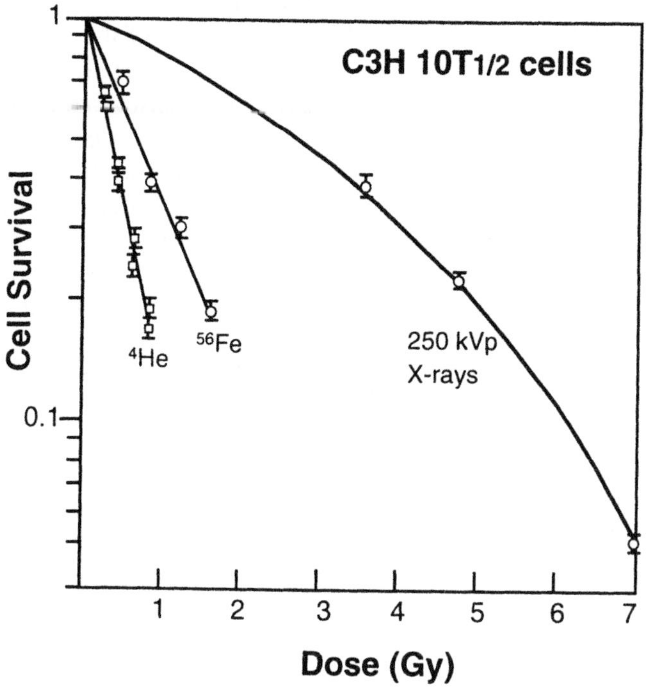

Fig. 1. Cell survival for C3H 10T1/2 cells exposed to X rays, high energy ^{56}Fe ions and ^{4}He ions. Error bars represent ±1 standard deviation.

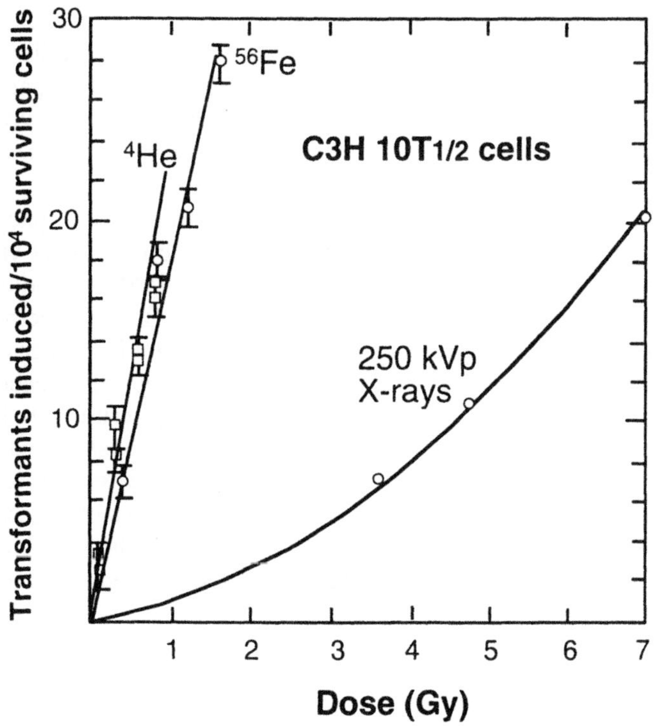

Fig. 2. Oncogenic transformation for C3H 10T1/2 cells exposed to X rays, ^{56}Fe ions and ^{4}He ions. Error bars represent ±1 standard deviation.

Oncogenic transformation for C3H 10T1/2 cells exposed to 250 kVp X rays, 150 keV/μm ^{4}He nuclei and 1 GeV/nucleon ^{56}Fe are shown in Figure 2. Data for direct comparison of the radiation types were fitted to the equation, $R_i = b + \alpha_i D + \beta D^2$ using a single common value of α for both radiations. In this equation, b is the background transformation rate, D is dose and subscript i refers to the radiation type. For each pair of radiation types, the residuals from the individual-α and the common-α fits were then compared using an F test, to examine the null hypothesis that the data from both radiation types came from the same distribution. In other words, instead of evaluating the significance of individual points, all data points were fit to a common α and residual errors (goodness of fit) were examined. It is apparent by this comprehensive method that oncogenic induction is quite different between the qualities of radiation. X-ray induction is curvilinear while ^{4}He and ^{56}Fe inductions are essentially linear. Comparison of the initial slopes for transformation induction results in RBE_{max} values of 22 and 16.5 respectively. In this series of experiments, the spontaneous transformation frequency was 0.65 x 10^{-4} transformants per surviving cell.

Clonogenic survival for C3H 10T1/2 cells exposed to 1 GeV/nucleon ^{56}Fe and X rays in the presence or absence of misoprostol was also measured. Misoprostol showed significant protection of cells exposed to X rays with a dose-reduction factor (DRF) of 1.2. However, misoprostol showed no ability to protect cells exposed to 1 GeV/nucleon ^{56}Fe nuclei (data not shown).

CONCLUSIONS

Extended manned space flights inevitably will result in significant doses to astronauts from high-LET radiation that includes high-energy ^{56}Fe nuclei. In this series of experiments, we examined the oncogenic potential of high-energy ^{56}Fe and compared heavy ion transformation induction with 250 kVp X rays and 150 keV/μm ^{4}He nuclei. Selection of 150 keV/μm ^{4}He nuclei is based on microdosimetric calculations that indicate the LET for 1 GeV/nucleon ^{56}Fe is ~143 keV/μm. However, it is already known that LET is not a good indicator of biological effect for heavy particles. Based on these studies, there is a significant difference for both cell lethality and oncogenic transformation between the two particles of similar LET. Indeed, several other investigators have made similar observations with cell survival and chromosomal aberrations as biological endpoints (Goodhead *et al.*, 1992; Goodwin *et al.*, 1996; Durante *et al.*, 1998).

Although exposure may be unavoidable, the biological effects from deep space exposures may be modified. Studies have demonstrated the ability of misoprostol to protect a wide variety of tissues from the toxic and oncogenic effects of X and gamma rays (Hanson *et al.*, 1988). In addition, misoprostol has recently been shown to protect tissue from damage after exposure to fission-spectrum neutrons (Hanson and Grdina, 1991). We conclude from the present study that misoprostol is neither a cytoprotector nor an oncoprotector of cells exposed to high-energy ^{56}Fe ions. Although the mechanism of action of misoprostol is not fully known, some of its cytoprotective qualities may result from its ability to stimulate repair of radiation damage. Since exposure to high-LET radiation significantly reduces the cell's ability to repair complex damage, it would seem that misoprostol was unable to restore full repair function. However, the value of misoprostol as a radioprotector remains since studies by Hanson and Grdina (1991) have demonstrated that with high-LET fission-spectrum neutrons, the combination of amifostine (administered as WR1065) and misoprostol showed significant radioprotection compared to either agent alone.

ACKNOWLEDGMENTS

This work was supported by grants P41 RR-11623 and CA 49062 from the National Institutes of Health and a joint NCI/NASA grant 73946. The Radiological Research Accelerator Facility (RARAF) is an NIH Supported Resource Center.

REFERENCES

Durante, M., L. Cella, Y. Furusawa, K. George, G. Gialanella, G. Grossi, M. Publiese, M. Saito and T. C. Yang, The Effect of Track Structure on the Induction of Chromosomal Aberrations in Murine Cells, *Int. J. Radiat. Biol.*, **73,** 253-262 (1998).

Goodhead, D. T., M. Belli, A. J. Mill, D. A. Bance, L. A. Allen, S. C. Hall, F. Ianzini, G. Simone, D. L. Stevens, A. Stretch, M. A. Tabocchini and R. A. Wilkinson, Direct Comparison Between Protons and Alpha-Particles at the Same LET. I. Irradiation Methods and Inactivation of Synchronous V79, HeLa and C3H10T1/2 Cells, *Int. J. Radiat. Bio.*, **61,** 611-624 (1992).

Goodwin, E. H., S. M. Bailey, D. J. Chen and M. N. Cornforth, The Effect of Track Structure on Cell Inactivation and Chromosome Damage at a Constant LET of 120 keV/μm, *Adv. In Space Research*, **18,** 93-98 (1996).

Hanson, W. R. and D. J. Grdina, Misoprostol, a PGE_1 Analog, Protects Mice from Fission-Neutron Injury, *Radiat. Res.*, **128**, S12 (1991).

Hanson, W. R., K. A. Houseman, A. K. Nelson and P. W. Collins, Radiation Protection of the Murine Intestine by Misoprostol, a Prostaglandin E_1 Analogue, Given Alone or with WR2721, is Stereospecific, *Prostaglandins Leukotrienes and Essential Fatty Acids*, **32**, 101 (1988).

LaNasa, P., R. C. Miller, W. R. Hanson and E. J. Hall, Misoprostol-Induced Radioprotection of Oncogenic Transformation, *Radiat. Res.*, **29**, 273 (1994).

Miller, R. C., S. A. Marino, D. J. Brenner, S. G. Martin, M. Richards, G. Randers-Pehrson and E. J. Hall, The Biological Effectiveness of Radon-Progeny Alpha Particles. II. Oncogenic Transformation as a Function of Linear Energy Transfer, *Radiat. Res.*, **142**, 54 (1995).

Yang, T. C., L. M. Craise and C. A. Tobias, Neoplastic Transformation by Heavy Charged Particles, *Radiat. Res.*, **104**, S177 (1985).

Wu, H., M. Durante, K. George and T. C. Yang, Induction of Chromosome Aberrations in Human Cells by Charged Particles, *Radiat. Res.*, **148**, S102 (1997).

Zeitlin, C., L. Heilbronn and J. Miller, Detailed Characterization of the 1087 MeV/nucleon Iron-56 Beam Used for Radiobiology at the Alternating Gradient Synchrotron, *Radiat. Res.*, **149**, 560 (1998).

Pergamon

Adv. Space Res. Vol. 22, No. 12, pp. 1725–1732, 1998

Printed in Great Britain
0273-1177/98 $19.00 + 0.00

PII: S0273–1177(99)00038–1

NEOPLASTIC TRANSFORMATION OF HAMSTER EMBRYO CELLS BY HEAVY IONS

Z. Han[1], H. Suzuki[1], F. Suzuki[2], M. Suzuki[3], Y. Furusawa[3], T. Kato, Jr.[1], and M. Ikenaga[1]

[1] *Radiation Biology Center, Kyoto University, Yoshida-konoecho, Sakyo-ku, Kyoto 606-8501, Japan*
[2] *Research Institute for Nuclear Medicine and Biology, Hiroshima University, 1-2-3 Kasumi, Minami-ku, Hiroshima 734-8553, Japan*
[3] *Third Research Group, National Institute of Radiological Sciences, 4-9-1 Anagawa, Inage-ku, Chiba 263-8555, Japan*

ABSTRACT

We have studied the induction of morphological transformation of Syrian hamster embryo cells by low doses of heavy ions with different linear energy transfer (LET), ranging from 13 to 400 keV/μm. Exponentially growing cells were irradiated with ^{12}C or ^{28}Si ion beams generated by the Heavy Ion Medical Accelerator in Chiba (HIMAC), inoculated to culture dishes, and transformed colonies were identified when the cells were densely stacked and showed a crisscross pattern. Over the LET range examined, the frequency of transformation induced by the heavy ions increased sharply at very low doses no greater than 5 cGy. The relative biological effectiveness (RBE) of the heavy ions relative to 250 kVp X-rays showed an initial increase with LET, reaching a maximum value of about 7 at 100 keV/μm, and then decreased with the further increase in LET. Thus, we confirmed that high LET heavy ions are significantly more effective than X-rays for the induction of *in vitro* cell transformation.

INTRODUCTION

In assessing the biological effects of heavy charged particles with high linear energy transfer (LET), especially with regard to prediction of radiation risk in space environment, additional information is needed because of the uncertainty in risk estimation from heavy ions in space. High LET heavy ions are relatively abundant in space, and account for about half of the total dose equivalent among astronauts in low orbit space flights at an altitude of less than 1,000 km (Benton, 1986; Badhwar *et al.*, 1993). With the coming era of the International Space Station and a possible future manned Mars Mission, an increasing number of humans will make longer and longer journeys into space. Under such circumstances exposure to space radiation is unavoidable. In general, total radiation dose in space is in the low dose range, nevertherless, it raises the possibility that high LET space radiation at low dose-rate exposure may present a higher cancer risk than the present estimate (NCRP Report 98, 1989; Badhwar *et al.*, 1993) due to the inverse dose-rate effect (Hill *et al.*, 1982; Miller *et al.*, 1993). Radiation carcinogenesis is thought to be the dominant stochastic effect of radiation in space. However, basic data on tumorigenesis of heavy ions are still limited, particularly at the low dose range.

To accumulate more knowledge on biological effectiveness of heavy ions for tumor induction, we have examined in the present study the frequency of morphological transformation in Syrian hamster embryo cells exposed to accelerated heavy ions with different LETs. Relative biological effectiveness (RBE) for cell survival and transformation, as a function of LET, for both carbon and silicon ions was determined.

MATERIALS AND METHODS

Cells

Pregnant Syrian hamsters were purchased from Shimizu Laboratory Supplies Co., Kyoto, Japan. Primary Syrian hamster embryo (SHE) cells were prepared from 13- to 14-day old embryos as described previously (Borek *et al.*, 1978; Watanabe *et al.*, 1984). Cells derived from a litter of embryos were combined and stored under liquid nitrogen. Among several different preparation of the SHE cells, we selected a single cell lot that showed the highest sensitivity to transformaion by a low dose (0.5 Gy) of X-rays and used throughout the experiments. The cells were grown in Dulbecco's modified minimum essential medium containing 10% fetal bovine serum (Hyclone, Utah, USA) and maintained in an atmosphere of 5% CO_2 in air at 37℃.

Transformation and Survival Assay

The transformation assay used in this experiment has been described elsewhere (Borek *et al.*, 1978; Watanabe *et al.*, 1984). Briefly, primary cultures of SHE cells stored in liquid nitrogen were plated in 75 cm^2 (T75) tissue culture flasks and incubated at 37℃ for about 24 hr. The following day, cells were trypsinized, subcultured in 25 cm^2 (T25) tissue culture flasks at a density of 5×10^4 cells/flask and further incubated for 2 days. Cultures were exposed either to graded doses of heavy ions with different LET values or X-rays. The irradiated cells were trypsinized and inoculated at a density of 400 cells/dish into 60-mm dishes containing feeder layer cells. Feeder cultures were prepared by irradiating SHE cells with a 50 Gy of X-rays and seeded at 10^4 cells/dish one day prior to heavy ion irradiation. For each transformation experiment, 80 dishes were seeded per dose for each heavy ion studied at a defined LET. In addition, 150 dishes were used for the non-irradiated control cells. Cultures were incubated for 8-10 days to allow for colony formation before being fixed and stained with Giemsa solution. Morphologically altered colonies, characterized by densely stacked cells and a crisscross pattern particularly evident on the periphery of the colony, were identified and scored. Transformation frequency was determined by dividing the number of transformants by the total number of colonies counted. The frequency of spontaneous cell transformation was about 2×10^{-5} transformants per survivor.

For survival assay, irradiated cells were appropriately diluted to yield about 50 colonies/dish, and plated into 60-mm dishes containing feeder layer cells. The dishes were then incubated and processed similarly to the transformation assay. At least 5 dishes were used per dose point in the cell survival assay.

Irradiation

The 290 MeV/u carbon-12 ion and 490 MeV/u silicon-28 ion beams were generated by the Heavy Ion Medical Accelerator in Chiba (HIMAC) synchrotron at the National Institute of Radiological Sciences, Japan. Details of the heavy ion irradiation and dosimetry have been described elsewhere (Kanai *et al.*, 1997). In brief, the energy fluences of the heavy ion beams were measured using a plastic scintillator, and LET was measured using a proportional counter filled with a tissue equivalent gas. The thickness of the lucite absorber plate (range filter) in front of the cell sample was adjusted to obtain LET values of approximately 13, 50 or 100 keV/μm for the carbon ions, and 150, 240 or 400 keV/μm for the silicon ions. The irradiation dose at the sample position was determined by multiplying the fluence by the stopping power in water. The dose to the sample was controlled by a parallel plate ionization chamber. The dose rates of the heavy ions ranged from 4-20 cGy/min for low dose irradiation and 1-2 Gy/min for high dose irradiation. As a reference radiation, we used 250 kVp X-rays filtered through 1 mm Al at a dose rate of 0.5 Gy/min. All irradiations were carried out at room temperature.

RESULTS

Cell Survival

Figure 1 shows survival curves of SHE cells irradiated with carbon ions, silicon ions or X-rays, as measured by colony-forming abilities. Survival curves of X-rays and carbon ions with LET of 13 keV/μm had a small shoulder at the low dose range (0-1 Gy), and at higher doses, survival decreased almost exponentially with the dose. In contrast, survival curves became approximately exponential with the dose for heavy ions with LET of 50 keV/μm or more. For all the LET range studied (13-400 keV/μm), slopes of the survival curves of heavy ions were much steeper than that of X-rays, and showed the greatest decrease in survival for irradiation with 100 keV/μm of carbon ions. The RBE of heavy ions relative to X-rays at 10% survival level was 1.2, 1.8 and 2.6 for 13, 50 and 100 keV/μm carbon ions and 2.3, 1.9 and 1.4 for 150, 240 and 400 keV/μm silicon ions, respectively (Table 1).

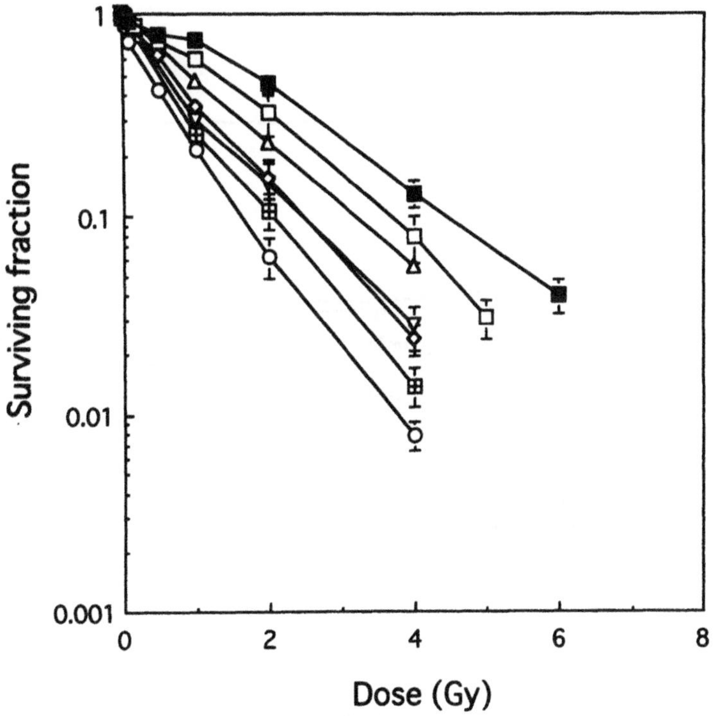

Fig. 1. Survival curves of Syrian hamster embryo (SHE) cells irradiated with heavy ions or X-rays. Carbon ions with LET of 13 keV/μm (□), 50 keV/μm (◇), 100 keV/μm (○), silicon ions with 150 keV/μm (⊞), 240 keV/μm (▽), 400 keV/μm (△), and 250 kVp X-rays (■). Each symbol represents the geometrical mean of four to seven independent experiments (see Table 1). Bars show standard errors if they are larger than the symbol.

Table 1. Relative Biological Effectiveness of Accelerated Heavy Ions for Cell Killing and Morphological Transformation in SHE Cells

Radiation	LET (keV/μm)	No. of experiments	Killing		Transformation	
			D_{10}[a] (Gy)	RBE	Slope[b] (10^{-4} cGy^{-1})	RBE
250 kVp X-rays	2.5	5	4.5	1.0	0.21	1.0
carbon ions	13	7	3.8	1.2	0.47	2.2
	50	7	2.5	1.8	0.90	4.3
	100	7	1.7	2.6	1.45	6.9
silicon ions	150	5	2.0	2.3	1.27	6.0
	240	4	2.3	1.9	0.82	3.9
	400	4	3.2	1.4	0.42	2.0

[a]Dose to give 10% cell survival.
[b]Slope of the linear regression curve for transformation induction.

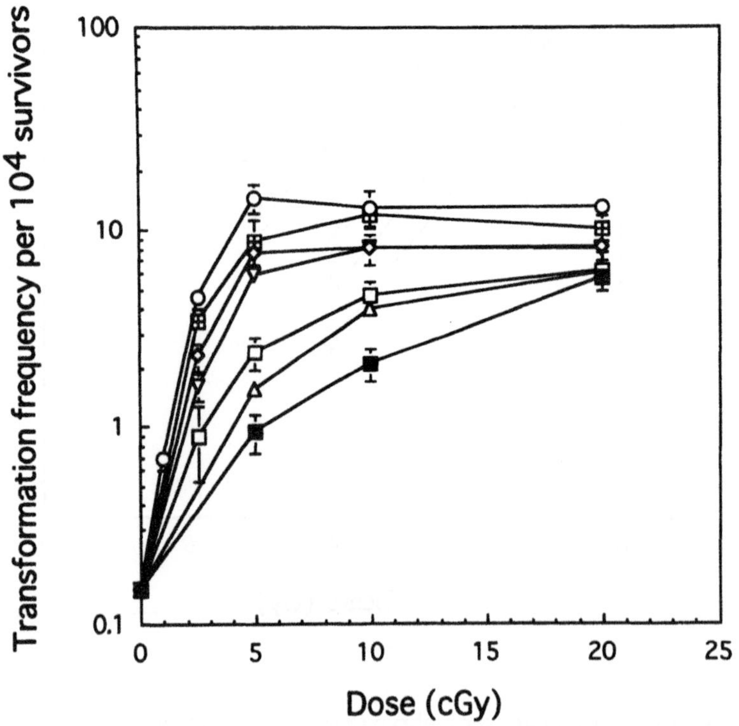

Fig. 2. Dose-response curves for morphological transformation of SHE cells after irradiation with either heavy ions or X-rays. Carbon ions at 13 keV/μm (□), 50 keV/μm (◇), 100 keV/μm (○), silicon ions at 150 keV/μm (⊞), 240 keV/μm (▽), 400 keV/μm (△), and 250 kVp X-rays (■). Results presented are the means and standard errors of four to seven independent determinations.

Morphological Transformation by Heavy Ions and X-rays

Figure 2 shows dose-response curves for the morphological transformation of SHE cells irradiated with carbon ions, silicon ions, and X-rays. The frequency of transformation induced by heavy ions increased sharply at doses from 0 to 10 cGy, and remained about the same level at higher doses up to 2 Gy (data not shown). Similarly, the transformation frequency for X-rays leveled off at doses over 0.5 Gy (data not shown). Therefore, only the data at doses no greater than 20 cGy are shown in Figure 2. The observed plateau level at very low doses (10-20 cGy) was markedly different from a previously reported dose-response in SHE cells which showed a plateau in the transformation yield at doses of 1-2 Gy (Suzuki, M. *et al.*, 1989). The difference may reflect the inherent difference in sensitivity between the two primary SHE preparations, since we specifically selected a lot which was highly responsive to low doses of X-rays. In any case, the transformation frequency per survivor for a given dose was consistently higher for heavy ions than for X-rays, and the highest frequency was observed for carbon ions with 100 keV/μm LET.

To calculate the RBE of heavy ions with a defined LET value, a linear model was fitted to the data for doses of 0 to 10 cGy, and regression coefficients were estimated by the least-squares method. The RBE of the heavy ions was determined as the ratio of the slope of the regression curve to that of X-rays (Table 1). The RBE for transformation increased with LET, reached a maximum of about 7 at 100 keV/μm, then decreased with the further increase in LET.

LET Dependence of RBE for Cell Killing and Transformation

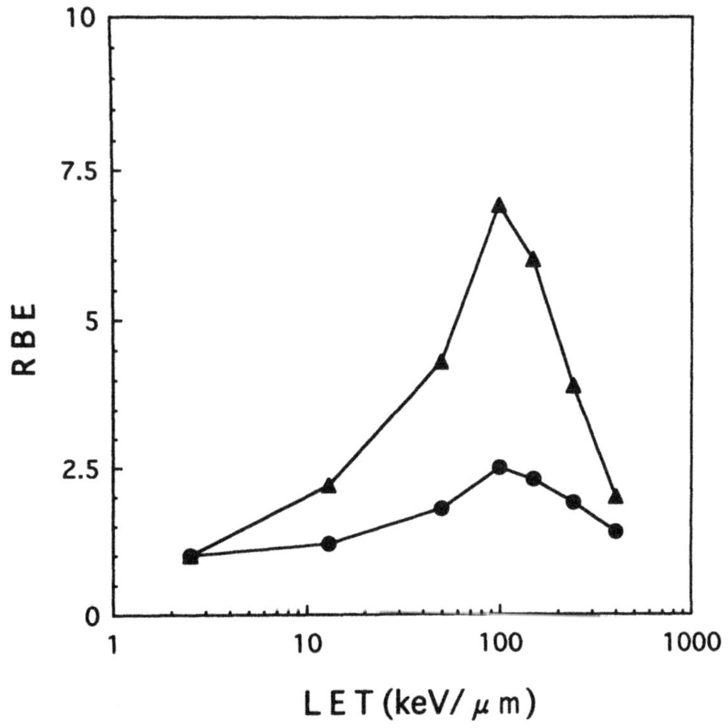

Fig. 3. Relative biological effectiveness (RBE) of cytotoxicity and transformation induction as a function of LET. Cell killing (●), transformation (▲).

Figure 3 shows the relationships between RBE and LET for both cell killing and morphological transformation with SHE cells. Both curves had a similar shape with a peak at 100 keV/μm LET, although the RBE value at any LET of the heavy ions was larger for transformation than for cell killing. All the data for the RBE versus LET relationships and actual RBE values are consistent with those reported previously for cell killing (Yang *et al.*, 1985; Hei *et al.*, 1988; Suzuki *et al.*, 1996; Sato and Soga, 1997) as well as transformation (Yang *et al.*, 1985, 1996; Suzuki, M. *et al.*, 1989).

DISCUSSION

Results of the present studies clearly showed that low doses of high LET heavy ions were more effective than X-rays for induction of morphological transformation of SHE cells. The RBE of heavy ions had a maximum value of 6-7 at LET of 100-150 keV/μm (Figure 3). These results agree with the previous reports on the LET dependence of RBE for cell transformation, which showed peak RBE values in the LET range of 100 to 150 keV/μm (Yang *et al.*, 1985; Hei *et al.*, 1988; Suzuki, M. *et al.*, 1989).

The relationship between RBE and LET for morphological transformation is very similar to that for cell lethality (Yang *et al.*, 1985; Suzuki *et al.*, 1996; Sato and Soga, 1997), 6-thioguanine-resistant mutation (Cox *et al.*, 1977; Tsuboi *et al.*, 1992; Suzuki *et al.*, 1996) and chromosome aberrations (Skarsgard *et al.*, 1967; Lloyd *et al.*, 1983; Suzuki *et al.*, 1996). For all these endpoints, the RBE first increases with LET, reaching a maximum at about 100 to 200 keV/μm , and then decreases with further increase in LET. For induction of cell killing and chromosome aberrations, the initial increase of RBE with LET is postulated to be the results of irreparable or long-persisting DNA double-strand breaks which are the responsible lesions for cell death and chromosomal anomalies (Lloyd *et al.*, 1983; Goodhead *et al.*, 1993), and their incidences increase with LET up to 200 keV/μm (Ritter *et al.*, 1977). It is generally thought that the rapid fall in RBE at LET above 200 keV/μm is a reflection of an excess amount of energy deposition in a target volume, i.e., energy in excess of that need to cause a single radiobiological event (Lloyd *et al.*, 1983; Tsuboi *et al.*, 1992; Goodhead *et al.*, 1993). The excess energy results in an "overkill" and a decrease in observed biological effect. Alternatively, the decrease of RBE at very high LET range might be due to an increase in surviving cells (cells not traversed by heavy ions) with increasing LET (Goodhead *et al.*, 1993; Sato and Soga, 1997).

Although we do not know the exact lesion(s) leading to cell transformation, DNA double-strand breaks appear to be the probable primary damage responsible for transformation (Yang *et al.*, 1996). This notion is supported by the following circumstantial evidence: (1) There is a good correlation between transformation and cell killing, deletion mutation, and chromosome aberrations, in terms of RBE versus LET relationships, (2) Yang *et al.* (1985, 1996) showed that potential transformation lesions produced by low LET radiation were repairable, whereas those induced by heavy ions with LET greater than 100 keV/μm were not, indicating that long-persisting DNA double-strand breaks are involved in transformation. It was also suggested that misrepair of the double-strand breaks could be a cause of cell transformation (Heilmann *et al.*, 1993; Yang *et al.*, 1996). In contrast, some investigators suggested that aneuploidy may have an important role in the neoplastic transformation of hamster embryo cells (Suzuki, K. *et al.*, 1989; Li *et al.*, 1997). The correlation between long-persisting DNA double-strand breaks and karyotype change in radiation-induced cell transformation is a subject for future study.

If DNA is the primary target, it is interesting to know how large is the target for cell transformation. The cross-section per unit fluence of heavy ions has been calculated by using equations, $S/S_0 = e^{-\sigma_i \Phi}$ for survival and $Ft = \sigma_t \Phi$ for transformation, where σ_i and σ_t are the inactivation and transformation cross-sections, respectively (Tsuboi *et al.*, 1992; Yang *et al.*, 1996). The transformation frequency, Ft, can be obtained from the slope of the regression curve shown in Table 1. Table 2 summarizes the cross-sections as

Table 2. Cross-section for Cell Inactivation and Transformation

LET (keV/μm)	Inactivation σ_i (μm^2)	Transformation σ_t (μm^2)
13	1.3	0.01
50	7.6	0.07
100	22.3	0.23
150	27.5	0.30
240	36.5	0.31
400	46.5	0.27

a function of LET. The inactivation cross-section increased with increasing LET up to about 100 keV/μm and reached a maximum value of about 50 μm^2 at 400 keV/μm, which is close to the geometric cross-section of the cell nucleus (Sato and Soga, 1997). In contrast, the transformation cross-section reached a plateau level at much lower LET range compared with inactivation cross-section, in agreement with previous observation (Yang *et al.*, 1996). The maximum transformation cross-section, about 0.3 μm^2, was one order of magnitude greater than that reported with C3H10T1/2 cells (Yang *et al.*, 1996), but a little smaller than the estimate for the same cells by other investigators (Hei *et al.*, 1988). In any case, the maximum transformation cross-section was about 100-fold greater than the mutation cross-section for 6-thioguanine resistance (Tsuboi *et al.*, 1992).

Apart from the nature of the potential transformation damage, the present data revealed the efficient induction of transformation at very small dose, as low as 1 cGy of 100 keV/μm carbon ions. In addition, heavy ions at dose as low as 2.5 cGy with LET range of 13 to 250 keV/μm induced transformation at a frequency 5 to 20 times higher than the controls (Figure 2). It is noteworthy that the 2.5 cGy of heavy ions with a quality factor (QF) of 20 corresponds to a dose equivalent of 500 mSv, a value similar to which astronauts are likely to be exposed during a 1 year stay at the International Space Station (Badhwar *et al.*, 1993).

ACKNOWLEDGMENTS

This work was supported by a Grant-in-Aid for Scientific Research on Priority Areas (No. 05278113) from the Ministry of Education, Science, Sports and Culture of Japan, and grants from the Japan Space Forum and the Heavy Ion Research Project of the National Institute of Radiological Sciences-Heavy Ion Medical Accelerator in Chiba (NIRS-HIMAC).

REFERENCES

Badhwar, G. D., A. C. Hardy, D. E. Robbins, and W. Atwell, Radiological Assessment for Space Station Freedom, *NASA Technical Memo.*, No. 104758 (1993).

Benton, E. V., Summary of Radiation Dosimetry Results on U.S. and Soviet Manned Spacecraft, *Adv. Space Res.*, **6(11)**, 315 (1986).

Borek, C., E. J. Hall, and H. H. Rossi, Malignant Transformation in Cultured Hamster Embryo Cells Produced by X-rays, 430-keV Monoenergetic Neutrons, and Heavy Ions, *Cancer Res.*, **38**, 2997 (1978).

Cox, R., J. Thacker, D. T. Goodhead, and R. J. Munson, Mutation and Inactivation of Mammalian Cells by Various Ionizing Radiations, *Nature*, **267**, 425 (1977).

Goodhead, D. T., J. Thacker, and R. Cox, Effects of Radiations of Different Qualities on Cells: Molecular Mechanisms of Damage and Repair, *Int. J. Radiat. Biol.*, **63**, 543 (1993).

Hei, T. K., K. Komatsu, E. J. Hall, and M. Zaider, Oncogenic Transformation by Charged Particles of Defined LET, *Carcinogenesis*, **5**, 747 (1988).

Heilmann, J., H. Rink, G. Taucher-Scholz, and G. Kraft, DNA Strand Break Induction and Rejoining and Cellular Recovery in Mammalian Cells after Heavy-ion Irradiation, *Radiat. Res.*, **135**, 46 (1993).

Hill, C. K., F. M. Buonaguro, C. P. Myers, A. Han, and M. M. Elkind, Fission-spectrum Neutrons at Reduced Dose Rates Enhance Neoplastic Transformation, *Nature*, **298**, 67 (1982).

Kanai, T., Y. Furusawa, K. Fukutsu, H. Itsukaichi, K. Eguchi-Kasai, and H. Ohara, Irradiation of Mixed Beam and Design of Spread-out Bragg Peak for Heavy-ion Radiotherapy, *Radiat. Res.*, **147**, 78 (1997).

Li, R., G. Yerganian, P. Duesberg, A. Kraemer, A. Willer, C. Rausch, and R. Hehlmann, Aneuploidy Correlated 100% with Chemical Transformation of Chinese Hamster Cells, *Proc. Natl. Acad. Sci. USA.*, **94**, 14506 (1997).

Lloyd, D. C., and A. A. Edwards, Chromosome Aberrations in Human Lymphocytes: Effect of Radiation Quality, Dose, and Dose Rate, in *Radiation-induced Chromosome Damage in Man*, edited by T. Ishihara and M. S. Sasaki, pp. 23-49, Alan R. Liss, New York, NY (1983).

Miller, R. C., G. Randers-Pehrson, L. Hieber, S. A. Marino, M. Richards, and E. J. Hall, The Inverse Dose-rate Effect for Oncogenic Transformation by Charged Particles is Dependent on Linear Energy Transfer, *Radiat. Res.*, **133**, 360 (1993).

NCRP, Guidance on Radiation Received in Space Activities, *Report 98*, National Council on Radiation Protection and Measurements, Washington, DC. (1989).

Ritter, M. A., J. E. Cleaver, and C. A. Tobias, High-LET Radiations Induce a Large Proportion of Non-rejoining DNA Breaks, *Nature*, **266**, 653 (1977).

Sato, Y., and F. Soga, Analysis of Relative Biological Effectiveness of High Energy Heavy Ions in Comparison to Experimental Data, *J. Radiat. Res.*, **38**, 103 (1997).

Skarsgard, L. D., B. A. Kihlman, L. Parker, C. M. Pujara, and S. Richardson, Survival, Chromosome Abnormalities, and Recovery in Heavy-ion- and X-irradiated Mammalian Cells, *Radiat. Res.*, **7**, 208 (1967).

Suzuki, K., F. Suzuki, M. Watanabe, and O. Nikaido, Multiple Nature of X-ray Induced Neoplastic Transformation in Golden Hamster Embryo Cells: Expression of Transformed Phenotypes and Stepwise Changes in Karyotypes, *Cancer Res.*, **49**, 2134 (1989).

Suzuki, M., M. Watanabe, T. Kanai, Y. Kase, F. Yatagai, T. Kato, and S. Matsubara, LET Dependence of Cell Death, Mutation Induction and Chromatin Damage in Human Cells Irradiated with Accelerated Carbon Ions, *Adv. Space Res.*, **18(1/2)**, 127 (1996).

Suzuki, M., M. Watanabe, K. Suzuki, K. Nakano, and I. Kaneko, Neoplastic Cell Transformation by Heavy Ions, Radiat. Res., **120**, 468 (1989).

Tsuboi, K., T. C. Yang, and D. J. Chen, Charged-particle Mutagenesis I. Cytotoxic and Mutagenic Effects of High-LET Charged Iron Particles on Human Skin Fibroblasts, *Radiat. Res.*, **129**, 171 (1992).

Watanabe, M., M. Horikawa, and O. Nikaido, Induction of Oncogenic Transformation by Low Doses of X Rays and Dose-rate Effects, *Radiat. Res.*, **98**, 274 (1984).

Yang, T. C., L. M. Craise, M. Mei, and C. A. Tobias, Neoplastic Cell Transformation by Heavy Charged Particles, *Radiat. Res.*, **104**, S177 (1985).

Yang, T. C., M. Mei, K. A. George, and L. M. Craise, DNA Damage and Repair in Oncogenic Transformation by Heavy Ion Radiation, *Adv. Space Res.*, **18(1/2)**, 149 (1996).

AUTHOR INDEX

CUMULATIVE INDEX FOR VOLUME 22